Physics and Chemistry in Space
Volume 6

Edited by
J. G. Roederer, Denver

Editorial Board:
H. Elsässer, Heidelberg · G. Elwert, Tübingen
L. G. Jacchia, Cambridge, Mass.
J. A. Jacobs, Edmonton
N. F. Ness, Greenbelt, Md. · W. Riedler, Graz

Siegfried J. Bauer

Physics of Planetary Ionospheres

With 89 Figures

Springer-Verlag

New York Heidelberg Berlin 1973

Siegfried J. Bauer

Associate Chief, Laboratory for Planetary Atmospheres, NASA Goddard Space Flight Center, Greenbelt, MD 20771/USA.

ISBN 0-387-06173-8 Springer-Verlag New York Heidelberg Berlin
ISBN 3-540-06173-8 Springer-Verlag Berlin Heidelberg New York

Monophoto typesetting and offset printing: Zechnersche Buchdruckerei, Speyer. Bookbinding: Konrad Triltsch, Würzburg.

Preface

Although there are, in addition to the classic but somewhat dated books*, some excellent recent books on ionospheric physics and aeronomy**, their scope is quite different from that of the present monograph.

This monograph concentrates on the fundamental physical and chemical processes in an idealized planetary ionosphere as a general abstraction, with actual planetary ionospheres representing special cases. Such an approach appears most appropriate for a concise introduction to the field, at a time when increasing experimental information on the ionospheres of other planets can be anticipated.

The main purpose of this monograph, in line with that of the whole series, is to appraise where we stand, what we know and what we still need to know. It is mainly addressed to graduate students and researchers who are in the process of getting acquainted with the field.

Within the scope of this monograph it would be impossible to do justice to all relevant publications. Hence, references are somewhat selective and largely limited to the more recent original papers and to authoritative reviews, the latter generally providing also detailed references regarding the historical development of the particular topic. Cgs (gaussian) units are generally used in this book, except where practical units are more appropriate.

This book has evolved from a graduate course of the same title which I gave at the Catholic University of America, Washington, D. C. in 1964/65 at the invitation of Prof. C. C. Chang. Although many of the details are new, the original organization of topics and the spirit of the course has been retained.

During the past decade I have benefited from the fruitful interaction with my colleagues at the Goddard Space Flight Center, first in the Planetary Ionospheres Branch and more recently in the Laboratory for

---

* cf. S. K. Mitra, *The Upper Atmosphere*, The Asiatic Society, Calcutta, 2nd ed 1952; J. A. Ratcliffe (ed), *Physics of the Upper Atmosphere* Academic Press, New York, 1960.

** cf. H. Rishbeth and O. K. Gariott. *Introduction to Ionospheric Physics*, Academic Press, New York, 1969; R. C. Whitten and I. G. Poppoff, *Fundamentals of Aeronomy*, John Wiley & Sons, New York, 1971; J. A. Ratcliffe, *An Introduction to the Ionosphere and the Magnetosphere*, Cambridge University Press, 1972.

Planetary Atmospheres. I am especially grateful to Dr. A. C. Aikin, Dr. R. F. Benson, Dr. S. Chandra, Dr. R. E. Hartle, Dr. J. R. Herman, Mr. J. E. Jackson and Dr. H. G. Mayr, who have read and commented on drafts of the manuscript. In addition, I want to thank Prof. S. H. Gross of the Graduate Center, Brooklyn Polytechnic Institute, Farmingdale, N. Y. and Dr. P. Stubbe of the Max-Planck-Institut für Aeronomie, Lindau (Harz), Germany, for their reading of the manuscript and their helpful comments. The responsibility for the views expressed in this book, however, remains my own.

Last, but not least, I wish to express my gratitude to Prof. O. Burkard of the University of Graz, Austria, for introducing me—two decades ago—to the field of ionospheric physics, and to my wife Inge and daughter Sonya for their patience with my preoccupation.

It is my hope that this monograph will be useful as an *introduction* to the physics of planetary ionospheres for a new generation of space scientists and as a *compendium* for the old.

April 1973   S. J. Bauer

Contents

Chapter V

Plasma Transport Processes

Chapter VI

Models of Planetary Ionospheres

Chapter VII

The Ionosphere as a Plasma

Chapter VIII

Experimental Techniques

Chapter IX

Observed Properties of Planetary Ionospheres

Introduction

The existence of the terrestrial ionosphere was verified experimentally by Appleton and Barnett in England and by Breit and Tuve in the U.S.A. less than half a century ago. An electrically conducting layer in the upper atmosphere had been postulated at the beginning of this century independently by Kennelly in the U.S.A., and by Heaviside in Great Britain to explain Marconi's pioneering transatlantic radio transmissions. Even earlier, during the 19th century, C. F. Gauss, W. Thomson (Lord Kelvin) and Balfour Stewart had argued for the existence of such a conducting layer to account for the daily variations of the Earth's magnetic field. Originally known as the Kennelly-Heaviside layer, E. V. Appleton, (who received the 1947 Nobel Prize in Physics for his pioneering work in ionospheric research), called this layer the "E-layer" for the electric vector of the reflected wave. When he found additional reflecting layers, in 1925, he named them correspondingly D and F layers, a terminology which is in present use. R. Watson-Watt, the pioneer of Radar, is responsible for naming the complex of reflecting layers the *ionosphere*.

Following the early theoretical work of E. O. Hulburt and Sydney Chapman, the contributions by D. R. Bates, V. C. Ferraro, D. F. Martyn, M. Nicolet, J. A. Ratcliffe and many others during the past three decades have led to our present understanding of the ionosphere as the result of the interaction between ionizing radiations from the sun and the neutral atmosphere, chemical processes and mass transport of plasma.

Over the past four decades the terrestrial ionosphere has been explored, first by radio methods from the ground, and since the advent of the space age also by means of spacecraft, thus making ionospheric physics a "space science". In addition, a new and powerful groundbased method, the incoherent backscatter radar technique was developed in recent years. Especially during the last decade, our knowledge of the terrestrial ionosphere has increased tremendously, in large part as the result of simultaneous observations of pertinent parameters of the neutral atmosphere, the ionizing radiations and the resulting charged particle distributions, their composition and temperature. This has led to a thorough examination of the physical and chemical processes governing the behavior of the ionosphere.

While the ionosphere of the Earth has been explored for many years, the existence of other planetary ionospheres was, until recently, primarily a matter of conjecture. However, it is a truism that any planet (or planetary satellite (moon)) that has an atmosphere also has an ionosphere, because of the presence of ionizing radiations in space. The first experimental evidence for other planetary ionospheres was obtained only during the past five years by means of a radio occultation experiment performed with the tracking and telemetry system onboard of the U.S. "Mariner" planetary probes. These measurements provided the initial experimental data for the ionospheres of our neighbor planets, Mars and Venus. *We shall define a planetary ionosphere as that part of the atmosphere of a planet, where free electrons and ions of thermal energy exist under the control of the force field (gravity, magnetic field) of the planet.*

The existence of an ionosphere results from the interaction of ionizing radiations with neutral constituents forming electron-ion pairs which ultimately recombine chemically. The ionosphere is maintained by a balance of ion pair production, chemical loss mechanisms and transport processes which can play the role of local sources or sinks according to the equation of continuity

$$\frac{\partial N}{\partial t} = q(n, \Phi) - L(N) - \mathbf{V} \cdot (N\boldsymbol{v})$$

where N is the plasma (electron, ion) density, q is the ion production rate depending on the neutral constituent density n and the ionizing radiation flux Φ, $L(N)$ is the chemical loss rate and $\mathbf{V} \cdot (N\boldsymbol{v})$ is the divergence of a flux $F = N\boldsymbol{v}$ due to mass transport with velocity $\boldsymbol{v}$.

Thus, for an understanding of the behavior of a planetary ionosphere we have to consider the properties of the neutral atmosphere, the ionizing radiations and the various physical and chemical processes. These will be discussed in the following chapters.

Chapter I

Neutral Atmospheres

I.1. Nomenclature

The atmosphere of a planet can be divided into a number of regions. It is quite natural that the terminology which has evolved for the terrestrial atmosphere should also be applied to other planetary atmospheres.* The terminology which is now generally accepted is based on the vertical distribution of temperature and composition, i. e., the parameters necessary for describing an atmosphere for which the hydrostatic condition can be used.

In terms of the temperature variation, the atmosphere can be divided as follows (see Fig. 1): The lowermost part of the atmosphere is the *troposphere*, where the primary heat source is the planetary surface** and heat is convected by turbulent motion. This leads to a convective or adiabatic density distribution. The vertical temperature gradient $\partial T/\partial z$ or adiabatic lapse rate is given by $\partial T/\partial z = -g/c_p$, where g is the acceleration of gravity and c_p is the specific heat at constant pressure. It is therefore dependent on the planet's acceleration of gravity and the atmospheric composition. (For Earth, the theoretical adiabatic lapse rate is $-10\,°K/km$, whereas the actually observed one is $-6.5\,°K/km$ due to the presence of water vapor and large-scale planetary motions.) The troposphere terminates at the *tropopause*, the level at which the temperature decrease with an adiabatic lapse rate ceases. Above the tropopause the temperature distribution is governed by radiative rather than convective processes and the temperature decreases much more slowly ($|\partial T/\partial z| < g/c_p$) or becomes essentially constant ($\partial T/\partial z \simeq 0$). For Earth the tropopause occurs at an altitude of 13 ± 5 km. Above the actual tropopause is the *stratosphere*. This region was originally thought to be isothermal throughout. In the terrestrial stratosphere the temperature, after being initially constant, increases with altitude due to

---

* An excellent introductory account of the fundamental properties of planetary atmospheres can be found in *Atmospheres* by R. M. Goody and J. C. G. Walker (Prentice-Hall Inc., 1972).

** This may not be the case for Venus.

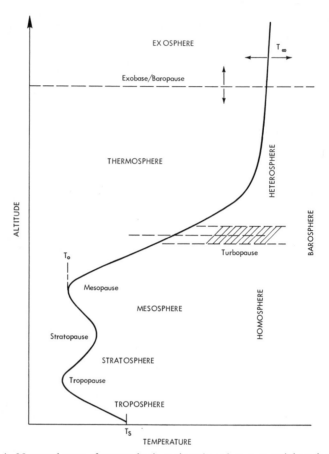

Fig. 1. Nomenclature of atmospheric regions based on terrestrial analogies

the ultraviolet (UV) absorption by ozone (O_3), reaching a temperature maximum at the *stratopause* (50 ± 5 km). Above this level the *mesosphere* begins, where $\partial T/\partial z < 0$, reaching a temperature minimum at the *mesopause* (Earth: 85 ± 5 km), due to the presence of CO_2 und H_2O which provide a heat sink by radiating in the infrared. The mesosphere is relatively unstable dynamically since convection is still prevalent.

Above the mesopause, EUV★ radiation is absorbed and in part used for heating, leading to a positive temperature gradient, $\partial T/\partial z > 0$. This region is called the *thermosphere*. In the lower thermosphere convection is the principal process of heat transport, while in the upper thermosphere

---

★ This term refers to extreme ultraviolet radiation $\lambda \lesssim 1800$ Å; when X-rays are included, the term XUV is used.

heat is transported by conduction, leading to an isothermal region ($T =$ const). The start of the isothermal region is termed the *thermopause*. In this region, the mean free path becomes large and collisions become negligible, so that light atmospheric constituents, whose velocity exceeds the gravitational escape velocity v_∞, can escape. This region is called the *exosphere*; the isothermal temperature above the thermopause is therefore also referred to as exospheric temperature. The exosphere begins at an altitude which has originally been called the critical level, but which is now generally referred to as *exobase*. This level is defined by the condition that the mean free path is equal to the local scale height, i.e., the logarithmic decrement of pressure with altitude. The exobase may also be called the *baropause*, since the entire atmosphere below that level is also referred to as the *barosphere*, i.e., the region where the barometric law holds. In the exosphere the velocity distribution is non-Maxwellian due to the escape of the high velocity particles and the density does not strictly follow the barometric formula, but has to be derived by considering the individual ballistic components of the atmospheric gas.

The troposphere and stratosphere together are also referred to (in a less precise way) as the *lower atmosphere*, while the regions above the stratosphere are called the *upper atmosphere*. (It should be noted, that without a stratospheric heat source, such as O_3 in the case of Earth, a planetary atmosphere may not possess a stratopause and the region above the tropopause may be called *either* the stratosphere *or* the mesosphere.) The lower atmosphere is the principal domain of *meteorology* whereas the upper atmosphere is that of *aeronomy*. However, it would be a mistake to rigidly compartmentalize the atmosphere, since interactions between the various regions do occur, although they are not yet fully explored. For the discussion of a planetary ionosphere, the emphasis must be, however, on the upper atmosphere.

The atmosphere can also be divided in terms of its composition into the *homosphere* and the *heterosphere*. These terms are somewhat less frequently used than the nomenclature based on the temperature distribution. In the *homosphere* the composition is uniform (Earth: 78% N_2, 21% O_2, 1% A plus minor constituents.) This region is characterized by turbulent mixing. In the *heterosphere*, the composition is varying due to dissociation of molecular constituents and as the result of diffusive separation. The level where diffusion rather than turbulent mixing becomes the controlling process is generally called the *turbopause*. More precisely, the turbopause can be defined as the level where the eddy diffusion (mixing) coefficient is equal to the molecular diffusion coefficient. Since these coefficients are different for different constituents, they will have different turbopause levels. In the terrestrial atmosphere, turbopause levels occur at altitudes 100 ± 10 km. The turbopause concentrations of atmospheric constituents represents important boundary conditions for their distribution in the upper atmosphere and the ionosphere.

The *ionosphere* is that region of the upper atmosphere where charged particles (electrons and ions) of thermal energy are present, which are the result of ionization of the neutral atmospheric constituents by electromagnetic and corpuscular radiation. The lower boundary of the ionosphere (which is by no means sharp) coincides with the region where the most penetrating radiation (generally, cosmic rays) produce free electron and ion pairs in numbers sufficient to affect the propagation of radio waves (D-region). The upper boundary of the ionosphere is directly or indirectly the result of the interaction of the solar wind* with the planet. For weakly or essentially non-magnetic planets (e. g.; Venus), the interaction region between the solar wind and the ionospheric plasma represents the termination of the ionosphere on the sunward side; it may be called the *ionopause*. On the nightside the ionosphere can extend to greater distances in a tail-like formation, representing the solar wind shadow. In the tail the extent of the ionosphere is limited by the condition for ion escape. For magnetic planets (Earth, Jupiter), the ionosphere terminates within the *magnetosphere* which comprises all charged particles of low (thermal) and high energies (radiation belts). In this case the solar wind interacts with the intrinsic planetary magnetic field terminating at the *magnetopause*. The termination of the ionosphere is then the indirect result of the solar wind interaction; e. g., in the case of Earth, represented by the boundary between solar-wind induced convective motions inside the magnetosphere and the corotating ionospheric plasma called the *plasmapause*. (The region inside the plasmapause is also called the *plasmasphere*; however, according to our definition this is simply part of the *ionosphere*.)

I.2. Barosphere: Distribution Laws

Static Atmosphere

The upper atmosphere, in fact the entire barosphere, can be characterized by its pressure and density distribution. The starting point for the derivation of this distribution is the hydrostatic equation

$$dp = -g\rho\,dz \qquad (1\text{--}1)$$

---

* The solar wind refers to the plasma which *continually* flows outward from the sun at supersonic speeds (~ 400 km/s) as the result of the expansion of the hot solar corona (cf. Vol. 5 of this series).

and the perfect gas law

$$p = n\ell T \qquad (1\text{--}2)$$

where p is the pressure, g is the acceleration of gravity, $\rho = nm$ is the mass density with n the number density, $m = \sum_j n_j m_j / n_j$ is the mean molecular mass, $\ell = 1.38 \times 10^{-16}$ erg/°K is the Boltzmann constant, T is the absolute temperature and z is the height variable.

Combining (1–1) and (1–2) and assuming first the simplest case of T, g and $m = \text{const}$ we obtain after integration

$$p = p_0 \exp\left(-\frac{mg}{\ell T} z\right) = p_0 \exp\left(-\frac{z}{H}\right)$$

$$n = n_0 \exp\left(-\frac{z}{H}\right) \qquad (1\text{--}3)$$

$$\rho = \rho_0 \exp\left(-\frac{z}{H}\right)$$

where $H = \ell T / mg$ is the atmospheric scale height.* Equations (1–3) represent the barometric formula or hydrostatic distribution in its simplest form. The barometric formula is a direct consequence of a Maxwellian velocity distribution.

The total content in a column of unit cross section of an atmosphere with constant scale height is given by

$$\mathscr{N} = \int_0^\infty n\,dz = n_0 H = \frac{p_0}{mg}. \qquad (1\text{--}4\text{a})$$

The total mass of a planetary atmosphere can be expressed by

$$M_{\text{atm}} = \left(\frac{p}{g}\right)_S 4\pi R_0^2 \qquad (1\text{--}4\text{b})$$

where the subscript S denotes the values of p and g at the planetary surface and R_0 is the planetary radius.

A more generally applicable form of the barometric law is given by the differential equation

$$\frac{dp}{p} = \frac{dn}{n} + \frac{dT}{T} = -\frac{dz}{H} \qquad (1\text{--}5)$$

which can be used throughout the barosphere.

It is often desirable to take into account the altitude variation of the acceleration of gravity in the barometric law, by defining a *potential*

---

* $H(\text{km}) \doteq 825\, T(°\text{K})/\hat{m}(\text{AMU})\, g(\text{cm sec}^{-2})$.

or *reduced* altitude z'. As long as the centrifugal force due to the planet's rotation is negligible, the acceleration of gravity is given by

$$g(z) = \frac{g_0 R_0^2}{(R_0 + z)^2} \qquad (1\text{--}6)$$

where R_0 is the planetary radius and g_0 is the acceleration of gravity at the planet's surface $(g_0 = GM/R_0^2$, with M the planetary mass and $G = 6.6695 \times 10^{-8}$ cm^3 g^{-1} sec^{-2}). Numerical values of g_0 for the planets are listed in Table 1. The reduced or potential altitude can then be defined by

$$z' = \int_0^z \frac{g(z)}{g_0} dz = R_0^2 \int_0^z \frac{dz}{(R_0 + z)^2} = \frac{R_0 z}{R_0 + z}. \qquad (1\text{--}7)$$

By using the reduced altitude, the variation of $g(z)$ is taken into account and g_0 can be used in the expression for the scale height, making H truly constant if $m, T = $ const.

Table 1

Planet	g_0 (cm/sec^2)
Mercury	370
Venus	880
Earth	980
Mars	370
Jupiter	2500
Saturn	1100
Uranus	950
Neptune	1160
Pluto	?

Taking into account (1–5) the atmospheric scale height is defined by

$$H = -\left(\frac{d\ln p}{dz}\right)^{-1} \qquad (1\text{--}8)$$

which is the inverse logarithmic decrement of pressure.

In the heterosphere and thermosphere the scale height varies with altitude as the result of the variation of m and T, leading to

$$\frac{dH}{dz} = \frac{1}{T}\frac{dT}{dz} - \frac{1}{m}\frac{dm}{dz} - \frac{1}{g}\frac{dg}{dz}. \qquad (1\text{--}9)$$

In this case a distinction must be made between atmospheric (pressure) scale height H and density scale height H_ρ, since $H \geqslant H_\rho$ [1]. (The two scale heights are the same only for an isothermal atmosphere with $m=$const.) This fact is important since the deceleration of artificial satellites resulting from atmospheric drag depends on the product $\rho H_\rho^{\frac{1}{2}}$.

The pressure and density scale heights are related by virtue of

$$H_\rho = -\left[\frac{1}{\rho}\frac{d\rho}{dz}\right]^{-1} = \frac{H}{1+\beta-2H/(R_0+z)} \qquad (1\text{--}10)$$

where $\beta=dH/dz$; the term $2H/(R_0+z)$ arises from the variation of g with height. The scale heights in units of 'reduced' altitude are thus given by

$$H'_\rho = \frac{H'}{1+\beta'} \qquad (1\text{--}11)$$

where $\beta'=dH'/dz'$.

Often the scale height gradient β can be assumed to be constant over certain height intervals. In this case the pressure and density distributions can be expressed by [2]

$$p = p_0\left(\frac{H}{H_0}\right)^{-\frac{1}{\beta}} = p_0\exp(-\zeta)$$

$$n = n_0\left(\frac{H}{H_0}\right)^{-\frac{(1+\beta)}{\beta}} = n_0\exp\left[-(1+\beta)\zeta\right] \qquad (1\text{--}12)$$

as the result of introducing a height variable ζ defined by $H=H_0\exp(\beta\zeta)$ or equivalently $d\zeta=dz/H$, if we take $\beta=dH/dz=$const, so that $H=H_0+\beta z$.

Any of the distributions given by (1–3) and (1–12) are called hydrostatic distributions. They are representative of a *mixing* distribution when the scale height H corresponds to a mean molecular mass $m=$const. When each individual constituent is distributed according to its own scale height $H_j=kT/m_jg$, such a distribution is also called a diffusive equilibrium distribution. In this case the various constituents obey Dalton's law of partial pressures.

Dynamic (Transport) Effects

The principal transport process in the upper atmosphere is *diffusion*. It is due to a gradient in relative concentration arising from *slight* deviations from a Maxwellian distribution. The relevant properties can be described by gas-kinetic methods of transport theory.

In considering the behavior of a minor constituent under diffusion, it can be assumed that the total pressure remains practically constant, while the minor constituent diffuses *upward or downward* due to the presence of local sources or sinks. This requires that the divergence of the flux $F_j(z)$ of the j-th constituent obeys the continuity equation

$$\frac{dF_j}{dz} = q_j(z) - L_j(z) \tag{1-13}$$

where q_j and L_j are the sources and sinks, respectively, for the j-th constituent.

The flux F_j which can be supported by a given density distribution $n_j(z)$ is given by [1, 3]

$$F_j = n_j w_j = -n_j D_j \left[\frac{1}{n_j} \frac{dn_j}{dz} + \frac{1}{H_j} + \frac{(1+\alpha_j)}{T} \frac{dT}{dz} \right]$$
$$- n_j K_D \left[\frac{1}{n_j} \frac{dn_j}{dz} + \frac{1}{H} + \frac{1}{T} \frac{dT}{dz} \right] \tag{1-14}$$

where w_j is the flow velocity, D_j is the molecular diffusion coefficient, H_j is the scale height of the j-th constituent, H is the (average) atmospheric scale height, α_j is the thermal diffusion factor and K_D is the eddy diffusion or mixing coefficient. Equation (1-14) contains the effects of molecular diffusion (first term) as well as eddy diffusion (second term) which is of principal importance below the turbopause, whereas molecular diffusion is predominant above the turbopause. Fig. 2 shows the effects of eddy and thermal diffusion.

The molecular diffusion coefficient in a multi-constituent gas is given by [4, 5]

$$\frac{1}{D_j} = \sum_{k \neq j} \frac{n_k}{b_{jk}} \tag{1-15}$$

where the b_{jk} are given by the binary encounters between particles of species j and k. Assuming rigid elastic spheres of diameters d,

$$b_{jk} = \frac{3}{8 d_{jk}^2} \left(\frac{\mathscr{k} T (m_j + m_k)}{2 \pi m_j m_k} \right)^{\frac{1}{2}} \tag{1-16}$$

where $d_{jk} = \frac{1}{2} (d_j + d_k)$.

Thus, the molecular diffusion coefficient is proportional to

$$D_j \propto T^{\frac{1}{2}} n^{-1}. \tag{1-17}$$

The thermal diffusion coefficient can also be estimated by assuming rigid elastic spheres. In the case of a minor constituent, $\alpha_j \rightarrow -\frac{5}{13}$. However, the rigid sphere model seems to overestimate α_j; a practical value for light minor constituents (H, D, He) is $\alpha_j = -0.25$ [5].

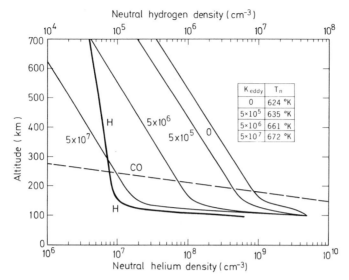

Fig. 2. Distribution of hydrogen and helium in a CO_2 atmosphere at $T=650\,°K$ representative of the Venus atmosphere. Helium distributions for different eddy diffusion coefficients K_D are given, and the effect of thermal diffusion is also included. The latter can be seen for the case $K_D=0$. The hydrogen distribution, because of strong upward diffusion supporting the escape flux, does not exhibit any substantial dependence on eddy or thermal diffusion. Note that below 200 km the scale heights of the light constituents are approximately that of CO_2 due to maximum upward flow, whereas at higher altitudes they approach that of a diffusive equilibrium. (After [95])

There is at present no completely satisfactory physical derivation of the eddy diffusion coefficient K_D★. For the Earth's atmosphere $K_D \simeq 5 \times 10^6$ cm^2 sec^{-1} has been deduced for steady state conditions [7, 8], while for the Martian CO_2 atmosphere an eddy diffusion coefficient two orders of magnitude greater than on Earth seems to be required to explain the scarcity of O, O_2 and CO in spite of the strong dissociation of CO_2 by solar radiation [8a].

For the case of a constant scale height gradient β, the diffusive flux above the turbopause can be expressed by

$$F_j = -\frac{D_j}{H}\left[\frac{dn_j}{d\zeta} + \left(\frac{m_j}{m} + \beta + \beta\alpha_j\right)n_j\right]. \qquad (1\text{–}18)$$

★ However, if it is assumed that K_D arises from turbulence produced by internal gravity waves (see p. 17) then

$$K_D = 1.4 \times 10^{-2}\, \tau_g^{-1} H^{-1} \lambda_x^4 \lambda_z^4 (\lambda_x^2 + \lambda_z^2)^{-\frac{5}{2}}$$

where λ_x and λ_z are the horizontal and vertical wavelength respectively, τ_g is the Brunt-Väisälä period, and H is the atmospheric scale height [6].

Integration of (1–18) with $F_j=$const leads to the density distribution under flow conditions [3]

$$n_j = n_{jo}\, e^{-A\zeta}\left[1 + \frac{B}{C}(e^{-C\zeta}-1)\right] \qquad (1\text{–}19)$$

where

$$A = \frac{m_j}{m} + \beta\alpha_j + \beta$$

$$B = \frac{F_j H_0}{n_{jo} D_0}$$

$$C = 1 - \left(\frac{m_j}{m} + \alpha_j\beta + \frac{\beta}{2}\right).$$

The parameters in B are taken at the lower reference level (subscript o) of the diffusion-controlled region, i. e., above the turbopause. The bracketed terms in (1–19) represent a correction factor to the non-flow (diffusive equilibrium) condition. The diffusive equilibrium distribution corresponds to the case $w_j=0$ (cf. 1–14). A more generally valid form of (1–19) for a variable flux $F_j(z)$ is given by [9]

$$n_j(z) = \overline{n_j(z)}\left[1 - \int\limits_0^z \frac{F_j(z)}{D_j(z)n_j(z)}\,dz\right] \qquad (1\text{–}20)$$

where $\overline{n_j(z)} = (T_0/T)n_{jo}\exp\left(-\int\limits_0^z dz/H_j\right)$ is the zero flow or diffusive equilibrium distribution.

A good approximation for the density under flow conditions, n_j can be obtained in terms of the integrated column density (total content) $\mathcal{N}$ of the main atmosphere (density n) through which the constituent j diffuses [10]. The flux F_j can be expressed in terms of the ratio of density under flow to the zero flow (static) density $\eta_j=n_j/\overline{n}_j$ according to

$$F_j = -D_j\overline{n}_j\frac{d\eta_j}{dz} = -\left(\frac{b_j\overline{n}_j}{n}\right)\frac{d\eta_j}{dz}$$

where $b_j=D_jn$ is the diffusion coefficient at unit concentration, with n the density of the main atmosphere. Using $n=-d\mathcal{N}/dz$ we can write

$$\frac{d\eta_j}{dz} = \frac{F_j}{b_j\overline{n}_j}\frac{d\mathcal{N}}{dz}$$

$$\eta_j = \frac{F_j\mathcal{N}}{b_j\overline{n}_j} + \eta_\infty$$

where η_∞ is the ratio of actual to static density at great heights and the density under flow follows as

$$n_j = \eta_j \overline{n}_j = \frac{F_j \mathcal{N}}{b_j} + \eta_\infty \overline{n}_j .$$

The flow which can be supported by diffusion in a given atmosphere without external sources or sinks $(\mathbf{V} \cdot \mathbf{F}_j = 0)$, is limited by condition $B = C$ in (1–19) and is given by

$$F_j^* = \frac{n_{j0} D_0}{H_0} \left[1 - \frac{m_j}{m} - \beta \left(\alpha_j + \frac{1}{2} \right) \right]. \qquad (1\text{--}21)$$

This leads to a density distribution at the reference level $(z=0)$, $d(\ln n_j)/dz = -(1 + \beta/2)/H$. This characteristic slope is approached at lower levels for near maximum flow, e. g., $0.95\, F_j^*$ and therefore the transition from a mixing distribution,

$$\frac{d(\ln n_j)}{dz} = -\frac{(1+\beta)}{H}$$

is almost indistinguishable. This is illustrated in Fig. 3 for hydrogen in N_2. For a constant scale height it can easily be shown that the

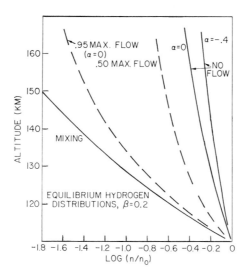

Fig. 3. Density distribution for mixing, diffusive equilibrium and flow conditions of the minor constituent hydrogen in a nitrogen atmosphere, having a scale height gradient $\beta = 0.2$, including the effect of thermal diffusion. (After Mange [3])

maximum upward flux

$$F_j^* = \frac{n_0 D_0}{H}\left[1 - \frac{m_j}{m}\right]$$

leads to a density distribution $n_j = n_{j0}\exp(-z/H)$, i. e., a distribution corresponding to the main *mixed* atmosphere through which the minor constituent diffuses. The density distribution of a minor constituent under flow and diffusive equilibrium for $H_j = \text{const}$ is illustrated in Fig. 4.

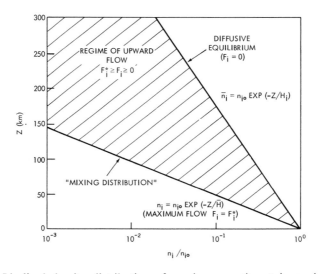

Fig. 4. Idealized density distribution of a minor constituent in an isothermal atmosphere under mixing, diffusive equilibrium and flow conditions

The time constant for (molecular) diffusion⋆, i. e., the characteristic time for attaining a diffusive equilibrium distribution is given by [3]

$$\tau_{Dj} \cong \frac{H^2}{D_j} = \frac{H^2}{b_j}n(z). \tag{1–22a}$$

For a constant scale height, τ_D decreases exponentially with altitude since it is proportional to $n(z)$.

⋆ This can be derived from the continuity equation $\partial n_j/\partial t = -\partial F_j/\partial z$, noting that $\tau \equiv ((1/n)(\partial n/\partial t))^{-1}$.

Similarly, the time constant for turbulent mixing *(eddy diffusion)* is given by

$$\tau_{K_D} \cong \frac{H^2}{K_D} \qquad (1\text{-}22\,\text{b})$$

where K_D is the eddy diffusion coefficient. (The altitude where $\tau_D = \tau_{K_D}$ is usually identified with the "turbopause".)

The equation of motion of the neutral gas (such as a wind system set up by pressure gradients) can be expressed by [11]

$$\frac{\partial \boldsymbol{v}_n}{\partial t} + \frac{\rho_i}{\rho_n} v_{in}(\boldsymbol{v}_n - \boldsymbol{v}_i) = \boldsymbol{g} - \frac{1}{\rho_n}\nabla p_n + 2(\boldsymbol{v}_n \times \Omega) + \frac{\eta}{\rho_n}\nabla^2 \boldsymbol{v}_n \qquad (1\text{-}23)$$

where $\boldsymbol{v}_n$ is the (vector) velocity of the neutral gas, $\boldsymbol{v}_i$ is the velocity of the ion gas (which may be constrained by a planetary magnetic field), v_{in} is the ion-neutral collision frequency, $\rho_{n,(i)}$ is the neutral (ion) mass density, $\boldsymbol{v}_n \times \Omega$ is the Coriolis acceleration, with Ω the angular velocity of the planet's rotation and η the coefficient of viscosity (η/ρ_n is the kinematic viscosity). The second term on the left hand side is referred to as the *ion drag* term, which arises from the interaction between the neutral atmosphere and the ionosphere. Attempts have been made to deduce the wind-system in the terrestrial thermosphere arising from the diurnal pressure field [11]. Fig. 5 shows the thermospheric wind field derived from (1-23). The atmospheric motion (winds) must also satisfy the conservation of mass through the equation of continuity

$$\frac{\partial \rho}{\partial t} = \nabla \cdot (\rho \boldsymbol{v})$$

for the neutrals and simultaneously for the ions. Accordingly, vertical motions must occur which balance the divergence or convergence resulting from the horizontal wind field.

Atmospheric disturbances can propagate from the lower to the upper atmosphere where they may be responsible for oscillations in the neutral gas as well as in the ionospheric plasma in form of *travelling ionospheric disturbances (TID)* [12].

Oscillatory solutions can be obtained from the equations of motion and continuity for the neutral gas (1-23), having the form

$$\frac{\Delta p}{p} \propto \frac{\Delta \rho}{\rho} \propto v_n \propto \exp i(\omega t - k r) \qquad (1\text{-}24\,\text{a})$$

where $k = (k_x + k_z)^{\frac{1}{2}}$ is the complex wave number, with $k_{x,z} = 2\pi/\lambda_{x,z}$.

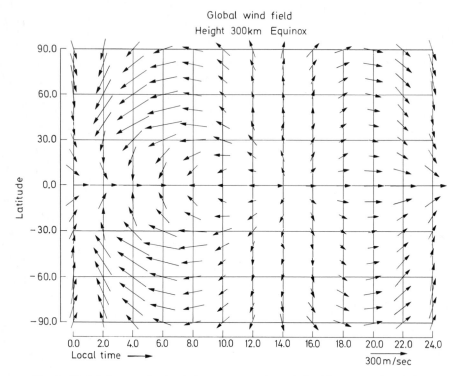

Fig. 5. Global thermospheric wind system at equinox derived from equation (1–23) using model values of $\nabla p_n/\rho_n$ based on satellite drag data and ion drag based on a semi-empirical model ionosphere (maximum wind speed ~ 300 m/sec) (Courtesy of P. Blum and I. Harris)

For *upward* propagation in a lossless atmosphere one obtains the solution

$$\frac{\varDelta\rho}{\rho} \propto \exp\left(\frac{\gamma g z}{2 c_s^2}\right) \equiv \exp\left(\frac{z}{2H}\right) \tag{1–24 b}$$

since the atmospheric scale height H and the sound speed $c_s = (\gamma\mathcal{k}T/m)^{\frac{1}{2}}$ are related by $H = c_s^2/\gamma g$, where γ is the ratio of the specific heats. This result is a direct consequence of the condition that the energy flux remain constant, i.e.,

$$\tfrac{1}{2}\rho v_n^2 = \text{const}$$

since in an isothermal atmosphere $\rho \propto \exp(-z/H)$.

Waves for which such propagation can take are determined by the following cutoff conditions: Waves whose angular frequency $\omega > \omega_a$

$=\gamma g/2c_s$ are termed *acoustic waves*, whereas those for which $\omega < \omega_g$ $=(\gamma-1)^{\frac{1}{2}} g/c_s$ are termed *internal gravity waves*. The cutoff frequency ω_a is called the *acoustic low frequency cutoff*, while the high frequency cutoff for internal gravity waves ω_g is called the (isothermal) *Brunt-Väisälä frequency*. Internal gravity waves can be considered a type of low frequency acoustic wave. (Tidal waves are essentially internal gravity waves where the effect of the Coriolis force is included and which propagate primarily horizontally.) These waves can be produced from energy supplied by tidal forces, from large-scale (planetary) wind systems, or as the result of atmospheric heating by energetic particles as occurs in the auroral zone.

In reality, upward propagation of internal gravity waves for a given ω is limited by the kinematic viscosity of the atmosphere as well as by reflections from a thermal barrier (i.e., where $dT/dz > 0$) so that only certain wavelengths are able to penetrate to higher altitudes [13].

I.3. Thermosphere—Thermal Structure

The energy budget of the upper atmosphere of planets is primarily governed by the heating of the gas due to the absorption of solar extreme ultraviolet (EUV) radiation by atmospheric constituents, by heat transport due to conduction and convection and by heat loss due to emissions in the infrared (IR) [14, 15]. Radiative loss by IR occurs when atmospheric constituents are present which have transition levels in the infrared. This is the case for atomic oxygen in the terrestrial atmosphere and H_2, CO and CO_2 in other planetary atmospheres.

In addition to solar EUV there are other possible sources of heating of the upper atmosphere of planets. Among them are collisions between charged particles and neutral constituents, Joule heating, conversion of dynamic energy into heat, such as from tidal motions and internal gravity waves, absorption of hydromagnetic waves and either direct or indirect heating by the solar wind. Direct solar wind heating can occur in planetary atmospheres without a significant screening magnetic field (magnetosphere) as on Venus, while indirect effects can occur via magnetospheric processes which are again induced by the solar wind (e.g., heating associated with auroras).

The thermal balance in the thermosphere is governed by a heat conduction equation of the form [11]

$$\rho c_v \left[\frac{\partial T}{\partial t} + v_n \cdot \nabla T \right] + p \nabla \cdot v_n - \nabla \cdot (K_n \nabla T) = Q_n - L_{IR} \qquad (1\text{-}25)$$

where ρ is the atmospheric mass density, c_v is the corresponding specific heat at constant volume, v_n is the velocity and p the pressure of the neutral atmosphere given by (1–23); K_n is the thermal conductivity which is a function of temperature according to $K_n = K_0 T^s$; Q_n represents the total heat production rate, which may be the sum of a number of processes, including recombination heating and heating by collisions with charged particles of the form $Q_{in} = A(T_e - T_n) + B(T_i - T_n)$ where A and B depend on the appropriate collision frequencies between electrons and neutrals and ions and neutrals, respectively. L_{IR}, the radiative loss term, depends on the number density of the emitting state which is related to the total number of the emitting species n_X by a partition function $f(T)$ and is thus of the form [14]

$$L_{IR} = n_X f(T);$$

the details have to be derived from radiative transfer considerations [15].

   A simplification of (1–25) has been considered by a number of workers in some detail, having the form [16]

$$\rho c_p \left(\frac{\partial T}{\partial t} + v_{nz} \frac{\partial T}{\partial z} \right) - \frac{\partial}{\partial z} \left(K_n(T) \frac{\partial T}{\partial z} \right) = Q_n - L_{IR}$$

together with a diffusive equilibrium distribution for n. For the one-dimensional case (vertical variability z only) the condition $dp/dt = 0$ allows replacement of c_v by c_p and the expression of v_{nz} as the velocity of a parcel of air identical to the vertical velocity of the isobaric surface ("breathing velocity" of the atmosphere) given by

$$v_B = T_n(z) \cdot \int_0^z \frac{1}{T_n^2} \frac{\partial T}{\partial t} \, dz \, .$$

Actually, v_{nz} must be considered the sum of the breathing velocity and a divergence velocity arising from the horizontal wind system by virtue of continuity [17].

   The principal heat source for the thermosphere is generally considered to be the conversion of absorbed solar EUV radiation into thermal energy by superelastic collisions. The heat production rate due to the absorption of solar EUV for a given wavelength and constituent is given by (see Chapter II)

$$Q_{uv} = \varepsilon_j n_j \sigma_a I_\infty e^{-\tau} \tag{1–26}$$

where ε_j is the fraction of absorbed UV energy which is transformed into thermal energy, i.e., the photoionization *heating efficiency*, which is thought to lie in the range from 0.3 to 0.6 and may also differ for different spectral ranges; I_∞ is the *intensity* (erg cm^{-2} sec^{-1}) at a given wavelength outside the atmosphere and is related to the *photon flux* Φ_∞ by

$$I_\infty = \left(\frac{hc}{\lambda}\right)\Phi_\infty$$

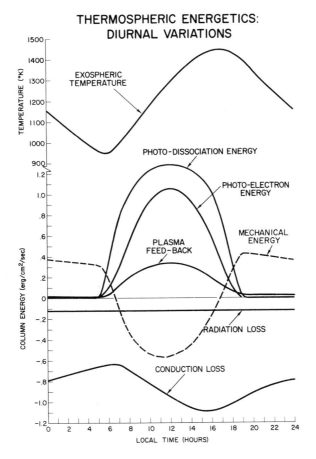

Fig. 6. Diurnal variation of exospheric temperature T_∞ and the integrated values (above 120 km) of the terms in the heat balance equation. "Photodissociation energy" represents the integrated heat input due to dissociation by the Schumann-Runge continuum; "Photoelectron energy" represents the heat input due to EUV photoelectrons; "Plasma feed-back" includes heat source due to collisions between the neutral and electron-ion gas and due to exothermic chemical reactions (dissociative recombination); "Mechanical energy" represents the integrated heat input due to the expansion and contraction ($\int p\nabla \cdot v_n \, dz$) of the atmosphere; "Radiation loss" represents the IR radiation loss above 120 km and "Conduction loss" represents the integral of the heat conduction term. (Courtesy of S. Chandra)

where h is Planck's constant, c the velocity of light; σ_a is the appropriate absorption cross section (typically in the range from 10^{-18} to 10^{-16} cm^2) and τ is the optical depth defined by

$$\tau = \int\limits_z^\infty \sec\chi \cdot \sigma_a \cdot n_j \, dz \qquad (1\text{--}27)$$

with χ the solar zenith angle*; for $\chi > 75°$, $\sec\chi$ must be replaced by the *Chapman function* Ch(χ) which allows for the curvature of the planet (cf. Chapter II).

The solution of the time-dependent heat balance equation (1–25) leads to a time variation of the temperature in the thermosphere, having a maximum which occurs later than that of the EUV heat input, due to the thermal inertia of the atmosphere. Fig. 6 illustrates the diurnal variation of the exospheric temperature T_∞ and that of the terms in the heat balance equation represented by their integrated values i.e., the total EUV heat input, the total IR loss, the total conduction loss and the mechanical energy $\int p\mathbf{V} \cdot v_n dz$, as obtained from a computer solution of (1–25). In addition to the EUV heat input, collisions between the neutral and the electron-ion gas can provide an additional heat source for the thermosphere according to (cf. 3–22)

$$Q_{coll} = 2\,N_i \frac{\mu_{in}^2}{m_i m_n} \, v_{in} \left[\frac{3}{2} k(T_i - T_n) + \frac{1}{2} m_n (v_i - v_n)^2 \right]$$

as can exothermic chemical reactions such as dissociative recombination (cf. Chapter IV)

$$X Y^+ + e \to X + Y + \Delta E.$$

Fig. 7 shows the vertical temperature profile at the diurnal maximum and minimum for the conditions shown in Fig. 6. A problem has arisen for the terrestrial thermosphere which has been termed the *diurnal phase anomaly* between atmospheric density and temperature. From satellite drag observation it was established [18] that the atmospheric density in the thermosphere reaches a maximum around 1400 LT and consequently a temperature maximum was inferred for the same time from atmospheric model considerations.** The solution of the heat

---

* $\cos\chi = \sin\varphi\sin\delta + \cos\varphi\cos\delta\cos(\Omega t)$, where φ is the latitude, δ is the solar declination, Ω is the atmospheric (planetary) rotation rate and t is time measured from the noon meridian.

** The magnitude of the diurnal phase anomaly is a function of atmospheric composition, since the individual atmospheric constituents attain their diurnal maxima at different times, e. g., He in the morning hours and O in the early afternoon. Thus, the lighter the constituent, the larger the phase anomaly.

conduction equation (1–25) on the other hand leads to a maximum in the temperature at about 1700 LT. A number of suggestions have been made to explain this phase discrepancy which range from the original assumption of a secondary ad hoc heat source [16] to the inclusion of dynamical effects and extension of the one-dimensional heat conduction to two and three dimensions [19, 20]. In addition, the ratio between temperature maximum and minimum also differs between theory and that derived from satellite drag observations, the latter being 1.3, whereas the former is ~ 1.5—2. Thermospheric temperature data inferred from groundbased incoherent scatter radar measurements suggest the peak temperature to occur near 1700 LT with a diurnal amplitude closer to the theoretical value. In the light of this result, a suggestion has been made [21] that the phase anomaly is due to the fact that the density and temperature at the turbopause is not time-independent but show a diurnal variation, whereas others [21a] view the phase anomaly as the result of wind-induced diffusion.

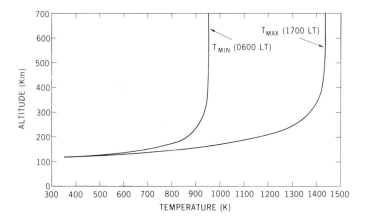

Fig. 7. Vertical temperature profiles in the thermosphere at the time of diurnal minimum and maximum for conditions in Fig. 6

A parametric study of the thermospheric heat balance equation of the type

$$\rho c_p \left\{ \frac{\Omega'}{\Omega} \frac{\partial T}{\partial t} + v_B \frac{\partial T}{\partial z} \right\} = Q + \frac{\partial}{\partial z} \left(K_n \frac{\partial T}{\partial z} \right) \qquad (1\text{–}28)$$

suggests that the diurnal variation of T_∞ responds to changes in the basic input parameters of equation (1–28) as shown in Table 2 [22].

Table 2. *Dependence of diurnal variation of exospheric temperature T_∞ on basic input parameters*

Cause	Effect		
Increase of:	t_{max}	$\dfrac{T_{min}}{T_0}$	$\dfrac{T_{max}}{T_{min}}$
Rotation rate of atmosphere, Ω' (superrotation)* Specific heat c_p	$\rightarrow$	$\uparrow$	$\curvearrowright$
Thermal conductivity K_n	$\leftarrow$	$\downarrow$	$\curvearrowright$
Absorption cross section, σ_a	$\leftarrow$	$\uparrow$	$\uparrow$
Pressure at base level, p_0	$\rightarrow$	$\uparrow$	$\curvearrowright$
Solar activity, I_∞	$\updownarrow$	$\uparrow$	$\updownarrow$

$\uparrow$ increase, $\downarrow$ decrease, $\curvearrowright$ goes through max, $\updownarrow$ ~const, $\leftarrow$ earlier, $\rightarrow$ later.

When a two-dimensional atmospheric model including winds is considered [23], the "diurnal amplitude" of temperature is found to be

$$\frac{T_{max}}{T_{min}} \propto \frac{Q(\Omega)}{|L_{con} - \Omega C|}$$

where $Q(\Omega)$ is the diurnal EUV heat source and $\Omega = 2\pi/d$ the angular frequency, with d the solar day, i.e., in general, Ω is the rotation rate of the planetary atmosphere; L_{con} is the heat loss due to conduction and convection, i.e., mass transport by winds, both assumed to be proportional to temperature and $C = c_v \rho$ is the heat capacity of the atmosphere. The time of the diurnal temperature maximum is related to these quantities by

$$t_{max} \propto \frac{C}{L_{con}}.$$

Empirical relations have been derived for the variation of exospheric temperature with solar activity and geomagnetic activity. As indicator of solar EUV intensity, the 10.7 cm solar radio flux (2800 MHz) observed on Earth has generally been employed, since continuous observations over long time intervals of the solar EUV radiation are still lacking. However, the close correlation between solar EUV and the 10.7 cm radio flux, $S_{10.7}$, has been established only over limited time intervals [24, 25]. Typical values of $S_{10.7}$ in units of 10^{-22} watts m^{-2} Hz^{-1} averaged over several solar rotations are: 75 for low solar activity (solar

---

* Super-rotation of planetary atmospheres may occur; i.e., the atmosphere (or parts of it) may rotate faster than the planet (Earth, Venus). (See H. Rishbeth, Revs. Geophys. Space Phys. 10, 799–819, 1972.)

minimum), 150 for medium activity and 230 for high solar activity (solar maximum).

Empirical relations between the terrestrial exospheric temperature derived from satellite drag and the 10.7 cm flux have been developed from observations extending over a full 11-year solar cycle [18]; a nearly linear correlation was found between T_∞ and $S_{10.7}$ (averaged over several solar rotations $\equiv 27$ days), according to which

$$\frac{\Delta T_\infty}{\Delta S_{10.7}} \simeq \frac{4\,^\circ\mathrm{K}}{10^{-22}\,\mathrm{watts\,m}^{-2}\,\mathrm{Hz}^{-1}}.$$

The temperature change due to magnetic activity is

$$\Delta T_\infty = 20\,K_p + 0.03\exp K_p$$

where K_p is the planetary magnetic index, which in turn is related to the solar wind velocity v_{sw}, according to

$$v_{sw}(\mathrm{km/sec}) \doteq 330 + 67.5\,K_p.$$

It should be noted that for planets without an appreciable intrinsic magnetic field, more direct solar wind heating may occur.

In the following we shall illustrate some of the general characteristics of planetary thermospheres which are the result of solar EUV heat input and heat conduction. For this purpose we consider a simple form of the heat balance equation neglecting the IR heat loss. For steady state conditions $(\partial T/\partial t = 0)$ we have

$$\frac{d\mathscr{F}}{dz} \equiv -\frac{d}{dz}\left(K_n\frac{dT}{dz}\right) = Q_{uv}$$

and by integration

$$\mathscr{F} = \int_z^\infty Q_{uv}\,dz = \varepsilon_j I_\infty(1 - e^{-\tau})$$

noting from the definition of optical depth that $dz = d\tau/n_j\sigma_a$. Since the heat flux $\mathscr{F} = -K_{nj}(T)dT/dz$, we can write⋆

$$\frac{dT}{dz} = \frac{\varepsilon_j I_\infty(1 - e^{-\tau})}{K_{nj}(T)}. \tag{1-29}$$

This represents the thermospheric temperature gradient which can be supported by the EUV heat input and heat conduction. In the exosphere, where the density is low, $\tau \to 0$ which leads to $dT/dz = 0$, i.e., an isothermal atmosphere. This constant temperature is also called the

---

⋆ We shall not use the subscript j for temperature.

exospheric temperature $T_\infty{}^\star$. The temperature gradient will have its maximum value where the optical depth is large ($\tau \to \infty$), i. e., in the lower thermosphere. This maximum temperature gradient is given by

$$\left(\frac{dT}{dz}\right)_{\text{max}} \cong \frac{\varepsilon_j I_\infty}{K_{nj}(T)}.$$

The thermospheric temperature gradient is therefore directly dependent on the solar EUV heat input (and the heat conductivity). For medium solar activity, the total flux at earth in the wavelength range 1310 to 270 Å (excluding the strong Lyman α line, 1216 Å) is about 2 erg cm^{-2} sec^{-1} (for further details see Chapter II).

The intensity for a given wavelength λ outside the planet's (P) atmosphere can be scaled from the terrestrial values according to $I_\infty^\lambda(\text{P}) = I_\infty^\lambda(\oplus)/D^2$, where D is the mean distance of the planet from the sun in astronomical units (AU).

Table 3. *Solar flux intensity outside a planetary atmosphere normalized to that observed in the terrestrial exosphere*

Planet	$\dfrac{I_\infty(\text{P})}{I_\infty(\oplus)}$
Mercury	6.65
Venus	1.9
Earth	1.0
Mars	0.43
Jupiter	3.7×10^{-2}
Saturn	1.1×10^{-2}
Uranus	2.7×10^{-3}
Neptune	1.1×10^{-3}
Pluto	0.65×10^{-3}

The time constant for equalization of a temperature difference by means of heat conduction is given by

$$\tau_{cond\,j} \cong \frac{n_j c_v \hbar L^2}{K_{nj}(T)} \qquad\qquad (1\text{--}30)$$

where L is a scale length over which this temperature difference exists.

By performing a second integration of the simple heat balance equation we can also determine the variation of the "exospheric" temperature T_∞ with the variation in solar EUV heat input over a solar cycle [26].

$\star$ A useful analytical model for the vertical temperature profile in the thermosphere is given by [14]

$$T(z) = T_\infty\{1 - a\exp(-bz)\} \quad \text{where } a = \frac{T_\infty - T_0}{T_\infty} \quad \text{and } b = \frac{1}{T_\infty - T_0}\left(\frac{dT}{dz}\right)_{z_0}.$$

The integrated form of (1–29) is given by

$$\int_{T_0}^{T_\infty} K_0 T^s dT = \int_{z_0}^{z_\infty} \varepsilon_j I_\infty (1 - e^{-\tau}) dz$$

where T_∞ is the exospheric temperature, T_0 is the temperature at the mesopause (which depends strongly on the IR loss) and z and z_0 are the corresponding altitudes. If we assume that $\tau \to 0$ at z_∞, i. e., in the exosphere, and $\tau \to \infty$ at z_0, i. e., at the base of the thermosphere, and allowing for $(n(z_\infty)\sigma_a)^{-1} \gg z - z_0$ (since the absorption cross section $\sigma_a \simeq 10^{-17}$ cm^2 and exospheric densities are of the order of 10^8 cm^{-3}) we obtain

$$K_0(T_\infty^{s+1} - T_0^{s+1}) \cong \frac{\varepsilon_j I_\infty}{n_j(z_\infty)\sigma_a}.$$

For $n_j(z_\infty)$ we can substitute the density at the base of the exosphere $n_c = (\sigma H)^{-1}$ (see next section), where $H = \mathscr{k} T_\infty / mg$ is the exospheric scale height, and σ is the gas-kinetic collision cross section. Thus we can write, neglecting T_0 compared to T_∞ (which seems permissible at least for the terrestrial planets)

$$T_\infty^s \propto \frac{\varepsilon_j T_\infty \mathscr{k} \sigma}{K_0 m_j g \sigma_a}$$

and

$$T_\infty \propto I_\infty^{\frac{1}{s}}.$$

The variation of T_∞ with solar EUV intensity is therefore determined by the temperature dependence of the thermal conductivity. It has been generally assumed, based on a rigid sphere approximation, that $K_{nj} = A_j T^{\frac{1}{2}}$. (Note that the heat conductivity is related to viscosity η by $K_n = \frac{5}{2}\eta c_v$, cf. [4].) However, experimental data [27] and quantum-mechanical calculations [28] show, that for atmospheric gases

$$K_{nj} = K_{0j} T^{s_j}$$

where $s_j > \frac{1}{2}$. The heat conductivity and its temperature dependence for likely constituents of planetary atmospheres is given in the following Table 4.

From satellite drag observations it is known that the terrestrial exospheric (daytime) temperature ranges from $\sim 800\,°$K at solar minimum to $\sim 2000\,°$K at solar maximum, i. e., a variation by a factor of 2.5, while the exact variation of the solar EUV intensity over a solar cycle has yet to be established. Since oxygen (O) is the principal constituent of the terrestrial thermosphere, according to Table 4, $T_\infty \propto I_\infty^{1/0.71}$ and we can infer from the temperature ratio of 2.5, that the EUV intensity should have changed by a factor of ~ 1.9.

Table 4. $K_{nj} = K_{0j} T^{s_j}$ (ergs cm^{-1} sec^{-1} °K^{-1})

Constituent	K_{0j}	s_j
N_2	27.21	0.80
O_2	18.64	0.84
CO_2	1.5	1.23
CO	26.16	0.80
O	67.1	0.71
He	20.92	0.75
H	16.36	0.73
A	18.3	0.80

For a CO_2 atmosphere (Mars, Venus) where $T_\infty \propto I_\infty^{1/1.23}$ holds, we obtain a temperature ratio over a solar cycle of ~ 1.7. This is in good agreement with observations by Mariner 4, 6 and 7, according to which the exospheric temperature of Mars ranges from $T_\infty \approx 300\,°K$ near solar minimum to $T_\infty \approx 500\,°K$ near solar maximum. For Venus, Mariner 5 measured $T_\infty \approx 650\,°K$ near the middle of the solar cycle. According to our estimate of the solar cycle variation, the exospheric temperature of Venus should range from $T_\infty \approx 470\,°K$ at solar minimum to $T_\infty \approx 800\,°K$ at solar maximum. These estimates are in good agreement with a more complete treatment of the heat balance equation [29].

It has been suggested that the condition for a gravitationally stable atmosphere given by the equality between r.m.s. thermal velocity $u_{rms} = (\overline{u^2})^{\frac{1}{2}} = (3 \mathscr{k} T_\infty / m_H)^{\frac{1}{2}}$ of hydrogen and its circular orbital velocity $v_{orb} = (g R)^{\frac{1}{2}}$, where g is the acceleration of gravity at the planetocentric distance R, determines the limiting exospheric temperature of a planet [30]. This limiting temperature is given by $T_\infty^* \approx m_H g R / 3 \mathscr{k}$ and leads to values of 400 °K for Mercury, 800 °K for Venus, 2000 °K for Earth, 300 °K for Mars and 25,000 °K for Jupiter. For Earth and Venus the exospheric temperatures throughout the solar cycle appear to lie below this limit, while for Mars this is the case only near solar minimum. The consequences of high exospheric temperature for the evaporation (escape) of gases from planetary atmospheres will be discussed in the following section.

I.4. Exosphere—Atmospheric Escape (Evaporation)

At any level of a planetary atmosphere there will be some particles which move upward with a velocity greater than that required for escape from the gravitational attraction of the planet. These particles are representative of the high velocity tail of the Maxwellian distribution. The actual escape, however, will depend on how frequent (or rather how infrequent) collisions are with other particles. At high enough altitudes the collision frequency will be low enough (i.e., the mean free path long enough) for escape to take place. This region is called the exosphere. The probability for a molecule traveling a distance z without a collision is given by

$$P(z) = e^{-\frac{z}{\lambda}}.$$

where λ is the mean free path

$$\lambda = (n\sigma)^{-1}$$

with n the total number density and σ the gas-kinetic collision cross section. Thus, the mean free path represents the distance for which this probability is $1/e$. The base of the exosphere z_c (exobase), originally called the critical level, is usually (but somewhat arbitrarily) defined by this probability requiring that

$$\int_{z_c}^{\infty} \frac{dz}{\lambda(z)} = \int_{z_c}^{\infty} n(z)\sigma\,dz = n_c H \sigma \equiv \frac{H}{\lambda} = 1. \tag{1–31}$$

Exospheric conditions, i. e., the possibility of evaporation, are therefore assumed to hold at altitudes where the mean free path is equal to or greater than the local scale height, $\lambda \geq H$. The density at the exobase is given by

$$n_c = (\sigma H)^{-1}. \tag{1–32}$$

This density value then defines the critical level or the exobase. Numerically, $n_c(\text{cm}^{-3}) = 2 \times 10^9 / H(\text{km})$ for $\sigma = 5 \times 10^{-15}\ \text{cm}^{-2}$ and the total content of an exosphere

$$\mathcal{N}_\infty \cong \sigma^{-1} = 2 \times 10^{14}\ \text{cm}^{-2}.$$

A "sharp" exobase obviously represents a highly idealized concept. Accordingly, it is assumed that below and at the critical level a Maxwellian distribution prevails, whereas in the exosphere the velocity distribution is truncated, due to absence of particles with speed above the escape velocity. Thus, the barometric law applies below the exobase, i.e., in the barosphere, but does not hold strictly in the exosphere*.

The minimum velocity for escape v_∞ is that for which the kinetic energy of the particle balances the potential energy in the gravitational field:

$$v_\infty = \left(\frac{2GM}{R}\right)^{\frac{1}{2}} = \sqrt{2gR} \tag{1–33}$$

---

* For this reason, the exobase is also called the baropause.

where G is the universal gravitational constant $G = 6.6695 \times 10^{-8} \, cm^3$ $g^{-1} \, sec^{-2}$, M is the mass of the planet, m that of the atmospheric particle and R is the radial distance from the center of the planet.

Table 5 lists the escape velocities for the planets for $R = R_0$; note that the escape velocity is $\sqrt{2}$ times the circular orbital velocity*.

Table 5

Planet	v_∞ (km/sec)
Mercury	4.3
Venus	10.4
Earth	11.2
Mars	5.1
Jupiter	60.0.
Saturn	36.0
Uranus	~ 22.0
Neptune	~ 22.0
Pluto	?

Our discussion of escape from planetary atmospheres is based on the classical treatment of Jeans [31], with extensions based on more recent work [32, 33, 34, 35], primarily in conjunction with the density distribution in a planetary exosphere. The basic assumptions of this treatment are: a) an isothermal atmosphere, b) a Maxwellian distribution even at the escape level.

The outward flux of particles whose velocity exceeds the local velocity of escape at a planetocentric distance R, i.e., the escape flux F_∞ (also called Jeans escape rate) is obtained by the product of the Maxwell-Boltzmann velocity distribution function and the vertical component of velocity $(v > v_\infty)$ over the upward hemisphere.

The escape flux relating to the critical level R_c (exobase) is given by the *Jeans* formula

$$F_\infty = \frac{u_0}{2\sqrt{\pi}} \cdot n_{jc}(1 + X_c)e^{-X_c} \qquad (1\text{--}34)$$

where n_{jc} is the density of the *escaping constituent* at the exobase**, $u_0 = (2 k T_\infty / m_j)^{\frac{1}{2}}$ is the most probable velocity at the temperature T_∞ and X is the escape parameter defined by

$$X(R) = \frac{G M m_j}{R k T_\infty} = \frac{v_\infty^2}{u_0^2}. \qquad (1\text{--}35)$$

---

* In the Russian literature the orbital velocity is referred to as "first cosmic velocity", while the escape velocity is called "second cosmic velocity".

** The exobase is defined by the *total* density at that level, $n_c = \sum_j n_{jc}$; cf. (1–32).

Thus, X_c is the escape parameter at the exobase. The escape flux at any level above the exobase can be expressed by

$$F_\infty = n_{jc}\langle v_R \rangle = \frac{u_0}{2\sqrt{\pi}} n_{jc}(1 + X_c)e^{-X_c} \cdot \left(\frac{X}{X_c}\right)^2$$

this flux decreases as R^{-2} to satisfy continuity[*]; $\langle v_R \rangle$ is the so-called *effusion velocity* which is defined by the ratio of escape flux to the concentration of the escaping constituent (Note that $\langle v_R \rangle$ is always subsonic).

The above expressions for the escape flux are based on the assumption of a Maxwellian velocity distribution at the exobase while escape itself leads to a perturbation of the Maxwellian distribution. A statistical (Monte Carlo) treatment however shows that the simple Jeans formula (1–34) overestimates the escape flux at worst by about 30%[**] [36, 37]. As long as the departure from a Maxwellian distribution is not too great ($X \gg 1$), X can also be expressed as a height parameter $X = R/H$. It should be noted that escape is important for $X < 15$, while for $X \le 1.5$ the exosphere becomes unstable and escape may become almost arbitrarily high [38]. The latter corresponds to the case when the thermal energy of the gas kinetic motion, $(3/2)\ell T$, exceeds the gravitational energy, i.e., $\overline{u^2}/v_\infty^2 \ge 1$. Typical values of the escape parameter X_c[***] for the lightest constituents of planetary atmospheres are listed in Table 6.

On a rotating planet, particles having velocities in the direction of the rotational motion, will more easily attain escape velocity than those moving in opposite direction. The resulting escape flux will therefore be primarily in the forward direction; the magnitude of the escape flux at the equator is also several times greater than that at high latitudes.

---

[*] The escape flux F_∞ is related to the maximum diffusive flux F_j^* of the escaping constituent in the barosphere by virtue of $R_l^2 F_j^* \equiv R_l^2 n_j w_j^* = R_c^2 F_\infty = n_{jc} R_c^2 \langle v_R \rangle$ where R_l is the lower boundary $(R_l < R_c)$; i.e.; n_{jc} depends on the diffusive flux in the barosphere. If $w_j^* < \langle v_R \rangle$, then an "escape bottleneck" occurs leading to an adjustment of density.

[**] With bulk flow at the exobase, however, the escape rate may be even higher than the Jeans escape rate [40].

[***] Useful formulas for the numerical evaluation of the escape parameter are

$$X \doteq 60\,\hat{m}\,[\text{AMU}] \cdot v_\infty^2(R)\frac{[\text{km/sec}]}{T_\infty[^\circ\text{K}]}$$

and for the most probable velocity

$$u_0 = 1.3 \times 10^4 \left(\frac{T_\infty}{\hat{m}}\right)^{\frac{1}{2}} [\text{cm/sec}].$$

Remember that $u_0 : \bar{u} : u_{\text{rms}} = \sqrt{2} : \sqrt{8/\pi} : \sqrt{3}$; mean square velocity $\overline{u^2} = \frac{3}{2}u_0^2$; $u_{\text{rms}} = (\overline{u^2})^{\frac{1}{2}}$.

Table 6. *Escape Parameter X_c*

| Planet | H | | | D | | He | |
	Sm*	SM**	Sm	SM	Sm		SM
Mercury		<2.4		<4.8		<9.6	
Venus	13	7	26	14	52		28
Earth	10	3	20	6	40		12
Mars	5	3	10	6	20		12
Jupiter		~1400					

The parameter governing rotational effects is given by

$$Y = \left(\frac{m_j \Omega^2 R^2}{2 \ell T_\infty} \right)^{\frac{1}{2}} \equiv \frac{\Omega R}{u_0} \tag{1-36}$$

where Ω is the angular rotation velocity.

The flux ratio between the pole and the equator for large Y is enhanced for increasing X. The difference between the fluxes at the pole is significant for different X values and is greater than that at the equator [39, 40]; the escape flux is always greater for smaller X (more efficient escape). The effect of rotation is illustrated in Fig. 8.

The concept of escape time is related to the available (total) column density for escape, $\mathscr{N}_\infty$. The ratio $\mathscr{N}_\infty/n_{jc}$, can be called the "escape length", corresponding to the length of a square centimeter column of density n_{jc} containing $\mathscr{N}_\infty$ particles [5]. The loss rate of $\mathscr{N}_\infty$ is given by

$$\frac{d\mathscr{N}_\infty}{dt} = -F_\infty = -\langle v_R \rangle n_{jc} = -\langle v_R \rangle \mathscr{N}_\infty \left(\frac{n_{jc}}{\mathscr{N}_\infty} \right). \tag{1-37}$$

Integration of (1-37) gives

$$\mathscr{N}_\infty(t) = \mathscr{N}_\infty(t=0) e^{-\frac{t}{\tau_J}} \tag{1-37a}$$

where the Jeans escape time

$$\tau_J = \frac{\mathscr{N}_\infty}{(n_{jc}\langle v_R \rangle)}. \tag{1-38}$$

Thus, τ_J represents a *time constant* and the fraction of the total content lost by escape is given by $(1 - e^{-t/\tau_J})$. In this connection it is worth remembering that the age of the planets is believed to be $t_0 \cong 4.5 \times 10^9$ yr $= 1.4 \times 10^{17}$ sec. If $\tau_J \ll t_0$, e.g., $t_0 = \underline{m}\tau_J$ with $\underline{m} > 7$, then τ_J can also be considered as the *time for* (virtually) *total escape*, since $e^{-t_0/\tau_J} \lesssim 10^{-3}$.

---

* Solar minimum conditions.
** Solar maximum conditions.

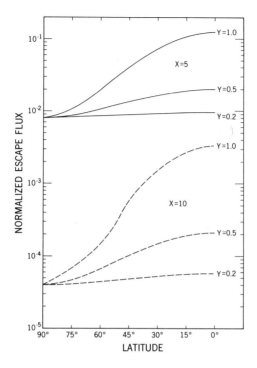

Fig. 8. Effect of planetary rotation on the escape flux as a function of latitude, for parametric values of the escape parameter $X = (v_\infty/u_0)^2$ and the rotational parameter $Y = \Omega R/u_0$. Normalized escape flux corresponds to $n_{jc}\langle v_R\rangle/u_0$. (Courtesy of R. E. Hartle)

Normally the big uncertainty lies in the estimate of $\mathcal{N}_\infty$, since the total content available for escape may include the constituent in non-escaping form, e. g., for hydrogen, water in an evaporating ocean.

Under the assumption that $\mathcal{N}_\infty \simeq n_{jc} \cdot H$, the escape time is given by [32]

$$\tau_J \cong \frac{1}{g}\left(\frac{2\pi \mathscr{k} T_\infty}{m_j}\right)^{\frac{1}{2}}\frac{e^{X_c}}{1+X_c}.\qquad (1\text{--}39)$$

Density Distribution in an Exosphere

In computing the density distribution in an exosphere, not only the escape component but also the particles in elliptic ballistic orbits i.e., those intersecting the exobase and in bound (satellite) orbits (not intersecting the exobase) must be considered. Because of the escape component, the velocity distribution in the exosphere departs from a Max-wellian and the barometric law ceases to describe the density distribution perfectly, although it can provide in some cases a good approximation to altitudes well above the exobase. The contribution of the various exospheric components is derived by determining the fraction of a Maxwellian distribution due to these components. The particle popula-tion in the exosphere can be divided into the following components (Fig. 9): 1) *ballistic orbits* $(v<v_{\infty})$; 2) *satellite (bound) orbits;* 3) *ballistic (escape) orbits* (3a) and *incoming capture orbits* (3b) $(v>v_{\infty})$ and 4) *hyperbolic orbits* not intersecting the exobase (interplanetary particles).

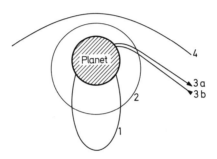

Fig. 9. Group of exospheric particles in different orbits: (1) ballistic orbits $v<v_{\infty}$; (2) bound satellite orbits; (3a) ballistic escape orbits and (3b) incoming capture orbits $(v>v_{\infty})$; and (4) hyperbolic orbits (interplanetary particles). (After John-son [35])

The proportion of particles falling into each group can be obtained by integrating the velocity distribution function over appropriate ranges, represented by a partition function ξ [2, 34, 35]. The density at any level R in the exosphere can be expressed by the product of the "barometric" density n_{jM} (based on a *complete* Maxwellian distribution) and the parti-tion function ξ, according to

$$n_j(R)=n_{jM}(R)\xi(X)=n_{jc}e^{-(X_c-X)}\cdot\xi(X). \qquad (1\text{--}40)$$

For no dynamical restriction of orbits $\xi=1$, $n_{jM}=\int\limits_{0}^{\infty} f(v)dv.$

The complete distribution of particles in the exosphere, normalized to the barometric density, is given by

$$\xi=\xi_1+\xi_2+\xi_3+\xi_4=1$$

where the subscripts refer to the components of the total particle popula-tion identified above and illustrated in Fig. 9.

Particles related to the critical level are represented by the partition functions [2, 34]

$$\xi_1 + \xi_3 = 1 - \frac{(X_c^2 - X^2)^{\frac{1}{2}}}{X} \exp\left\{\frac{X^2}{(X+X_c)}\right\} \qquad (1\text{-}41)$$

and particles not reaching the critical level (exobase) by

$$\xi_2 + \xi_4 = \frac{(X_c^2 - X^2)^{\frac{1}{2}}}{X} \exp\left\{-\frac{X^2}{(X+X_c)}\right\}. \qquad (1\text{-}42)$$

The elliptic components ($v < v_\infty$) can be expressed by

$$\xi_1 + \xi_2 = \Phi(X^{\frac{1}{2}}) + \tfrac{1}{2}\Phi''(X^{\frac{1}{2}}) \qquad (1\text{-}43)$$

where $\Phi(X)$ is the error function and Φ'' its second derivative, which are tabulated.

The hyperbolic components ($v > v_\infty$) expressed by

$$\xi_3 + \xi_4 = 1 - \Phi(X^{\frac{1}{2}}) - \tfrac{1}{2}\Phi''(X^{\frac{1}{2}}). \qquad (1\text{-}44)$$

The bound (satellite) component* is given by

$$\begin{aligned}\xi_2 = \frac{(X_c^2 - X^2)^{\frac{1}{2}}}{X_c} \exp\left\{-\frac{X^2}{(X+X_c)}\right\} \\ \cdot \left[\Phi\left(\frac{X}{(X+X_c)^{\frac{1}{2}}}\right) + \frac{1}{2}\Phi''\left(\frac{X}{(X+X_c)^{\frac{1}{2}}}\right)\right]\end{aligned} \qquad (1\text{-}45)$$

while the hyperbolic interplanetary component is given by

$$\begin{aligned}\xi_4 = \frac{(X_c^2 - X^2)^{\frac{1}{2}}}{X_c} \exp\left\{-\frac{X^2}{(X+X_c)}\right\} \\ \cdot \left[1 - \Phi\left(\frac{X}{(X+X_c)^{\frac{1}{2}}}\right) - \frac{1}{2}\Phi''\left(\frac{X}{(X+X_c)^{\frac{1}{2}}}\right)\right].\end{aligned} \qquad (1\text{-}46)$$

The above equations lead to the components *at the exobase*

$$\begin{aligned}\xi_1 &= \Phi(X_c^{\frac{1}{2}}) + \tfrac{1}{2}\Phi''(X_c^{\frac{1}{2}}) \quad \dots (v < v_\infty) \\ \xi_3 &= 1 - \Phi(X_c^{\frac{1}{2}}) - \tfrac{1}{2}\Phi''(X_c^{\frac{1}{2}}) \dots (v > v_\infty) \\ \xi_2 &= \xi_4 = 0 .\end{aligned} \qquad (1\text{-}47)$$

Eqns. (1–47) show that ξ_3 is practically negligible for a constituent whose escape rate is very small, i.e., which has a large value of X_c.

---

* The contribution of this component is still a subject of debate, since its presence depends on the rates at which particles enter and leave this group by various processes (e. g., radiation pressure; see G. E. Thomas and R. C. Bohlin, J. Geophys. Res. 77, 2754, 1972).

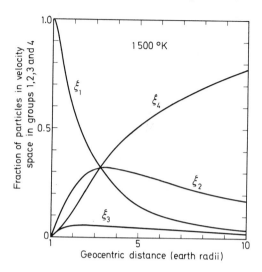

Fig. 10. Partition functions ξ for the group of particles shown in Fig. 9 as a function of geocentric distance R for $X = 4.65$ (i.e., H at 1500 °K). (After Johnson [35])

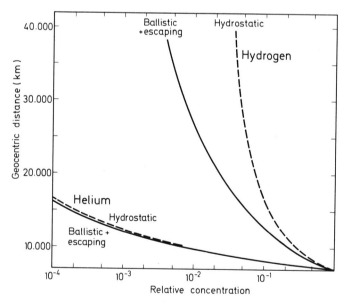

Fig. 11. Distribution of H and He in the terrestrial exosphere at 1700 °K showing the departure from a hydrostatic distribution for H due to escape. (After Nicolet [2])

Fig. 10 shows exospheric distributions of various components, represented by the partition functions ξ given above, for hydrogen in the Earth's exosphere at $T_\infty = 1500\,°K$ ($X_c = 4.65$). The relative concentration of terrestrial hydrogen and helium at $T_\infty = 1700\,°K$ are shown in Fig. 11 [2]. The hydrogen distribution based on the ballistic and escaping components departs strongly from the hydrostatic (barometric) law ($X_c = 4.35$), whereas the departure is almost non-existent for helium at the same temperature ($X_c = 17.4$).* The fraction of particles with velocity $v > v_\infty$ as a function of the escape parameter X is illustrated in Fig. 12.

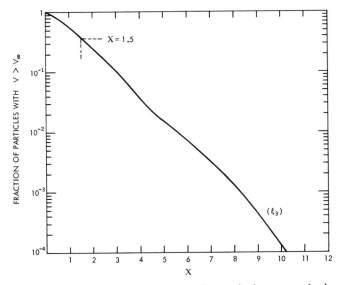

Fig. 12. Fraction of exospheric particles (ξ) whose velocity v exceeds the escape velocity v_∞, as function of the escape parameter X. The critical value of $X = 1.5$ represents Öpik's condition for atmospheric blow-off [38]

As the result of escape, the Maxwellian velocity distribution is truncated. Thus, at high altitudes the temperature at the exobase T_∞, is not representative. However, a kinetic temperature can be defined by the second moment of the velocity distribution function in the moving frame of reference, i.e., moving with velocity $\langle v_R \rangle$, according to

$$T_{\text{kin}} = \frac{2T_\infty}{3u_0^2}\left[\langle v^2 \rangle - \langle v_R \rangle^2\right]. \qquad (1\text{–}48)$$

---

* For a constituent subject to escape, the density *decreases* with increasing temperature, in contrast to constituents controlled by the barometric law (cf. Fig. 13).

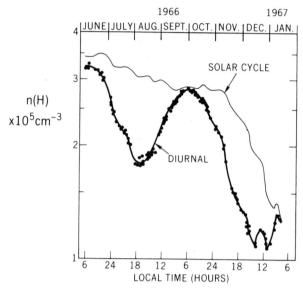

Fig. 13. Solar cycle and diurnal variation in terrestrial hydrogen as the result of escape, i.e., the hydrogen concentration decreases with increasing temperature. The hydrogen concentration corresponds to an altitude of 350 km and was derived from in-situ measurements of H^+ and O^+ together with a model of O, according to the chemical equilibrium relation $n(H) = (8/9)\, n(H^+) n(O)/n(O^+)$ (cf. (4-23)). The increase in solar activity during the observation period (1966/67) ($100 \leq S_{10.7} \leq 140$) and corresponding increase in exospheric temperature caused a decrease in H by a factor 2.5, whereas the diurnal variation corresponds to an increase in $n(H)$ by a factor 2 from day to night. (After H. C. Brinton and H. G. Mayr, [35a])

The ratio of the kinetic temperature to the temperature at the exobase for $X_c \gg 1$ is given by [34]

$$\frac{T_{\text{kin}}}{T_\infty} = \frac{\pi^{\frac{1}{2}}(3 + 2X_c)X^{\frac{1}{2}}}{16 X_c^2} + \frac{2}{5}X. \qquad (1\text{-}48\,\text{a})$$

When the mean energy is dominated by the escaping component, the ratio approaches [34]

$$\frac{T_{\text{kin}}}{T_\infty} \to 0.12. \qquad (1\text{-}48\,\text{b})$$

The kinetic temperature distribution for hydrogen (H) in the terrestrial exosphere according to (1-48) is shown in Fig. 14.

The density distribution discussed above (1-40) applies to a stationary (non-rotating) planet. A generalization of the velocity distribution for a corotating exobase, satisfying the collisionless Boltzmann equation, leads to a density distribution, in which the density at the equator exceeds that at the pole, for a given planetocentric distance [39]. Further

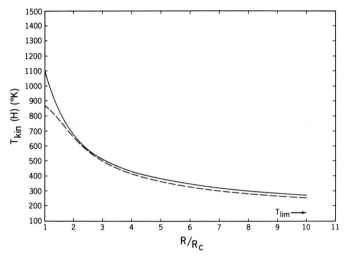

Fig. 14. Decrease of kinetic temperature of hydrogen in the exosphere resulting from escape, for hydrogen distributions based on the data in Fig. 13. Solid line refers to daytime, dashed line to night time situation. The limiting kinetic temperature under escape $T_{lim} = 0.12\,T_\infty$ is also indicated. (Courtesy of R. E. Hartle)

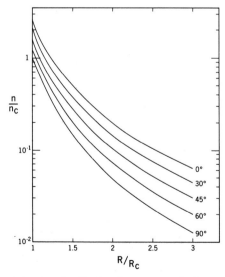

Fig. 15. Normalized density distribution for an exospheric constituent ($X = 4.65$, $Y = 1.0$) in a rotating exosphere as a function of latitude when the exobase is uniform. (Courtesy R. E. Hartle)

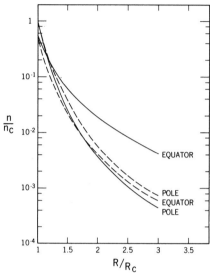

Fig. 16. Comparison of normalized density distribution for a non-rotating (dashed) exosphere with non-homogeneous exobase ($X=10$, $Y=0$) and a rotating exosphere (solid lines) for $X=10$ and $Y=1$. (Courtesy of R. E. Hartle)

refinements [40], including a non-homogeneous (in temperature and density) exobase, lead to even more drastic effects of rotation. Features of a rotating vs. a stationary exosphere are illustrated in Figs. 15 and 16. The effect of rotation becomes pronounced when the rotational parameter (1–36) $Y = \Omega R/u_0 \geq 1$, i. e., when the rotational velocity exceeds the most probable thermal velocity of the exospheric constituent.

The following Table 7 lists the values of the rotational parameter at the exobase Y_c, for important constituents in the exospheres of Venus, Earth, Mars and Jupiter. Since $Y_c \propto u_0^{-1}$, rotational effects for a given planet are most pronounced at low temperatures and for high mass numbers. The upper limits of Y_c in Table 7 are based on estimated exospheric temperatures for minimum solar activity.

Table 7

Y_c	H	He	O
Venus*	$\lesssim 0.06$	$\lesssim 0.12$	$\lesssim 0.24$
Earth	$\lesssim 0.15$	$\lesssim 0.3$	$\lesssim 0.6$
Mars	$\lesssim 0.13$	$\lesssim 0.26$	$\lesssim 0.52$
Jupiter	$\lesssim 7.5$	$\lesssim 15$	–

* Assuming 4-day atmospheric superrotation; otherwise negligible.

Loss of Atmospheric Particles by Non-thermal Processes

In addition to thermal escape, particles can be lost from an exosphere by ionization. This process will become important when the travel time of an escaping particle to a distance R is of the order of its life-time towards ionization τ_{ion}. This defines the distance R_{ion} beyond which loss by ionization will become the controlling factor [34]

$$\frac{R_{\mathrm{ion}}}{R_c} = \left[\frac{3 v_\infty(R) \tau_{\mathrm{ion}}}{2 R_c}\right]^{\frac{2}{3}}.$$

Photo-ionization, charge exchange with, and 'scavenging' by solar wind ions, especially for "non-magnetic" planets will cause a depletion of the planetary exosphere density distribution [40a] (see Chapter V). On the other hand, fast neutral particles due to solar wind proton charge exchange can penetrate to lower altitudes (cf., the return component of ξ_3). Non-thermal escape of neutrals can also occur as the result of kinetic energy gained by the neutrals in exothermic chemical reactions as long as $\Delta E \geq m v_\infty/2$ [41, 42]. This effect should be important for Mercury and Mars. Fig. 17 shows the escape energy as a function of

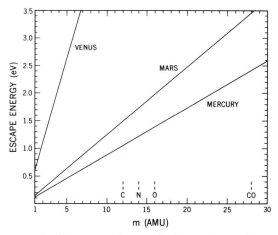

Fig. 17. Energy required for escape of atmospheric constituents from the exospheres of Mercury, Mars and Venus. This energy can be supplied by exothermic chemical reactions, for which $\Delta E \geq m v_\infty^2/2$

mass number for these planets and Venus. In addition to neutral gas escape, loss of atmospheric particles can also occur in the form of ions, since evaporation of ions (in the absence of a constraining magnetic field) can take place when conditions for an *ion-exosphere* are satisfied.

In fact, evaporating ions can reach supersonic velocities ("polar wind" on Earth) due to the presence of an accelerating electrostatic polarization field. Ion escape will be discussed in more detail in Chapter V.

I.5. Physical Properties of Planetary Atmospheres

It is now commonly accepted that the atmospheres of the terrestrial planets (Mercury, Venus, Earth and Mars) are of secondary origin, having lost their primordial constituents long time ago. This, however, is probably not the case for the Jovian Planets (Jupiter, Saturn, Uranus and Neptune). The secondary atmospheres are most likely due to outgassing of volatile materials from the planetary interior and radio-active decay products such as helium He^4 (from the decay of uranium and thorium) and argon (from potassium K^{40}), with the possible exception of oxygen in the Earth's atmosphere which seems to be related to the presence of life [43, 44]. (See Table 8 for constituents of planetary atmospheres.)

The Terrestrial Planets

There is at present no direct evidence that *Mercury*, the planet closest to the sun has an atmosphere. An upper limit of 0.01 mb of CO_2 has been inferred from groundbased spectroscopic observations. However, a very thin atmosphere, possibly an exosphere whose base is the planetary surface, may nevertheless be present [45]. Since the density of the planet is earth-like (although Mercury is much smaller in size), it may have a core and a crust. Therefore, internal radioactive decay may lead to outgassing of argon and helium and together with accretion of these constituents and of Ne^{20} from the solar wind, could be responsible for a tenuous atmosphere with a surface pressure in the range from 10^{-10} to 10^{-12} mb. The surface temperature of Mercury has been found to be $\sim 500\,°K$ averaged over the sunlit side.

In spite of their widely different atmospheres at the present time, it has been proposed that the main outgassing constituents from the planetary interiors have been the same for *Venus, Earth* and *Mars*, i. e., water vapor and CO_2 with a small amount ($<1\%$) of nitrogen. Their different evolutionary path has been suggested to be the consequence of the initial effective surface temperature due to their distance from the sun together with a "runaway" greenhouse effect on Venus due to H_2O and CO_2 [46]. It is now well established that Venus has a dense and hot atmosphere, while Mars has a thin and cold one relative to

Earth [47]. The surface pressures of Venus, Earth and Mars are in the approximate ratio 100:1:0.01, while their surface temperatures T_S are 750 °K, 300 °K and 250 °K, respectively. Both, Venus and Mars have CO_2 as their principal atmospheric constituent, whereas Earth has N_2 and O_2 [48]. The temperature and pressure distribution for Venus, Earth and Mars are shown in Figs. 18 and 19.

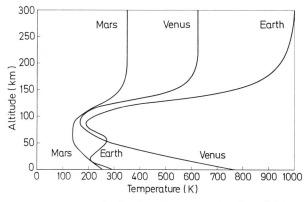

Fig. 18. Model temperature distributions for Venus, Earth and Mars consistent with our present understanding of these planetary atmospheres

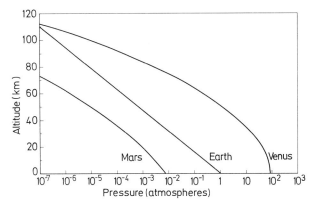

Fig. 19. Model pressure distributions for the atmospheres of Venus, Earth and Mars based on observations and theory

It has been suggested that because of the initial high surface temperature of Venus, H_2O if present at the formation of the planet, was always in the atmosphere in form of vapor, since the surface temperature was always above the boiling point of water at the given pressures in the early history of the planet. The complete absence of liquid water together with the high surface temperature then allowed a substantial accumulation of CO_2 in the atmosphere, since reactions between silicates in the crust and CO_2 proceed rapidly only in the presence of liquid water and are also strongly temperature dependent. The present lack of H_2O is explained as either due to photo-dissociation by solar ultraviolet radiation, with the hydrogen having escaped and the oxygen consumed by various oxidation processes at the surface or due to an initial absence of H_2O [48 a]. The composition of the Earth's atmosphere is thought to be due to the initial (lower) surface temperature allowing water to condense at the surface and accelerating the removal of CO_2 from the atmosphere and the accumulation of the inert gas, nitrogen, supplied from volcanic activity and oxygen associated with life. Mars, on the other hand, with its low initial temperature, caused the volcanic steam to freeze at the surface and allowed the accumulation of CO_2 in the atmosphere, but to a much lesser extent on account of its low temperature [46]. Although the above evolutionary path of the atmospheres of the terrestrial planets is largely speculative, it is reasonably consistent with present groundbased and spacecraft observations.

The physical parameters of Earth have, of course, been well established by observations. In addition to the principal atmospheric constituents and their dissociation products, the upper atmosphere of Earth contains helium (which is of radiogenic origin) and hydrogen (which is the result of photo-dissociation of water vapor and methane) as the major light constituents, in addition to atomic oxygen. Because of the range of exospheric temperatures over the solar cycle, the terrestrial hydrogen corona is variable. A hydrogen corona has also been observed for Mars and Venus [47]. Helium, though probably present, particularly on Venus, has yet to be identified directly. The presence of oxygen in both the Venus and Mars upper atmospheres has been established, but its concentration is extremely small compared to Earth, raising the interesting question of the stability of a CO_2 atmosphere. Since photo-dissociation should be very effective, competing processes of recombination must exist if the main heavy constituent of the upper atmospheres of Mars and Venus is indeed CO_2 as present observations seem to suggest. The CO_2 problem remains one of the most enigmatic ones regarding the atmospheres of Mars and Venus [49, 49 a].

Because of the relatively small scale heights the exobase for Mars and Venus occurs at a much lower altitude ($\lesssim 200$ km) than for Earth (~ 500 km). Due to the absence of a significant planetary magnetic field, the solar wind can therefore interact with the exospheres of Venus and Mars at rather low altitudes.

The Jovian Planets

In contrast to the terrestrial planets, the Jovian or outer planets have reducing atmospheres with large amounts of hydrogen, suggesting that they represent primordial atmospheres whose origin is similar to that of the solar system. According to present evidence [48, 50, 51], *Jupiter* and *Saturn* have atmospheres which in addition to hydrogen contain methane (CH_4) and ammonia (NH_3), while *Uranus* and *Neptune*, seem

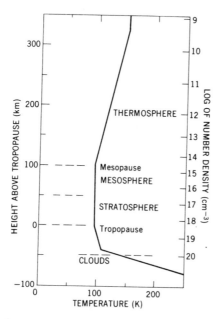

Fig. 20. Model temperature distribution as function of altitude and number density for the Jovian atmosphere. (After Hunten [259])

Table 8. *Gases Identified in Planetary Atmospheres*

Mercury	No definite identification (Ne, A, He?)
Venus	CO_2, CO, HCl, HF, H_2O, O, H, He?, (D?), (N_2?), NH_3
Earth	N_2, O_2, H_2O, A, CO_2, Ne, He, CH_4, K, N_2O, H_2, H, O, O_3, Xe
Mars	CO_2, CO, H_2O, O, H, (He?), (N_2?), O_3, O_2
Jupiter	H_2, (He), H, CH_4, NH_3
Saturn	H_2, CH_4; (He), NH_3?
Uranus	H_2, (He), CH_4
Neptune	H_2, (He), CH_4
Pluto	No identifications

to be richer in the heavier elements. Although helium is assumed to be a major constituent of the outer planets' atmospheres, experimental evidence for its presence will most likely have to come from spacecraft observations. As the result of their great distance from the sun, the exospheric temperatures of the outer planets is expected to be low ($\sim 150\,°K$ for Jupiter), unless there are other sources of heating such as dissipation of acoustic energy. Fig. 20 shows a model temperature distribution in the Jovian atmosphere. For *Pluto* no atmospheric constituents have been identified.

Chapter II

Sources of Ionization

The principal ionizing radiations responsible for the formation of planetary ionospheres are solar extreme ultraviolet (EUV) and X-rays (XUV radiation), and corpuscular radiation; i. e.; galactic cosmic rays and energetic particles having their origin in the solar system (solar cosmic ray protons, auroral particles, solar wind). In addition, interaction of meteors with atmospheric gases can lead to the formation of some ionization.

II.1. Solar EUV Radiation and X-Rays

The source of the solar EUV (1750 Å—170Å) and X radiation (170Å—1Å) is in the various layers of the solar "atmosphere". The following Table 9 lists the appropriate wavelengths at their point of origin in the solar atmosphere [52, 52a].

Table 9

Wavelength (Å)	Source of radiation
> 1800	Photosphere
1200—2000	Transition Photosphere-Chromosphere
900—1800	Chromosphere
100—1000, Lyα (1216)	Transition to Corona
10—200	Quiet Corona
5—100	Coronal active region
1—50	Thermal Radiation ⎱ Solar Flares
0.01—10	Nonthermal Burst ⎰

The absorption of solar EUV radiation and X-rays in a planetary atmosphere leads to *photoionization* and *photodissociation* of atmospheric constituents. The molecular constituents H_2O, CO_2 and O_2 can be dissociated by relatively long wavelength radiation in the *(Schumann-*

Runge continuum) range 1750—1300Å,* the latter according to

$$O_2 + h\nu\,(\lambda < 1750\,\mathring{A}) \to O(^3P) + O(^2D)$$
$$CO_2 + h\nu\,(\lambda < 1670\,\mathring{A}) \to CO + O(^1D).$$

Dissociation can also take place below the true dissociation limit due to excitation of molecules into a state which dissociates. This *pre-dissociation* occurs for O_2, CO_2 and H_2, but not significantly for N_2 accounting for its apparent stability against photodissociation. For H_2, radiation $\lambda < 850\,\mathring{A}$ leads to either dissociation or ionization.

The ionization thresholds (ionization potentials, IP) for common constituents of planetary atmosphere are listed in the following Table 10 (Note: $\lambda(\mathring{A}) = 12{,}395/IP(eV)$).

Table 10

Constituent	$\lambda(\mathring{A})$	IP (eV)
NO	1340	9.25
NH_3	1221	10.15
C	1100	11.3
O_2	1026	12.1
H_2O	985	12.60
CH_4	954	13.00
H	912	13.59
O	911	13.61
CO_2	899	13.79
CO	885	14.0
H_2	804	15.41
N_2	796	15.58
A	787	15.75
Ne	575	21.56
He	504	24.58

Accordingly, EUV radiation of wavelengths below the Lyman alpha line (1216Å) is primarily responsible for the formation of planetary ionospheres. However, trace constituents, of meteoric origin, can be ionized by even longer wavelengths. Typical examples are: Na (2410Å/ 5.14 eV); Ca (2028Å/6.11 eV); Mg (1622Å/7.64 eV); Fe (1575Å/7.87 eV); and Si (1520Å/8.15 eV).

Rocket and satellite observations during the past decade have led to a detailed identification of the solar emission line spectrum, while

---

* This wavelength range with an intensity of $\geq 2\,\text{erg cm}^{-2}\,\text{sec}^{-1}$ represents the predominant EUV heat source for the terrestrial thermosphere.

relatively accurate values for the intensities of these radiations have become available only rather recently [53, 54, 55]. Solar photon fluxes at Earth for moderate solar activity are summarized in Table 11 [55].*

Table 11

Wavelength (Å)	Φ_∞ (10^9 ph cm^{-2} sec^{-1})**	I_∞ (erg cm^{-2} sec^{-1})**
1215.7 (Ly α)	300	5
1027—911	11.61	0.23
(1025.7, Ly β)	(3.5)	(0.067)
(977, C III)	(4.4)	(0.090)
911—800	8.3	0.20
800—630	2.4	0.064
630—460	4.7	0.17
(584.3, He I)	(0.9)	(0.03)
460—370	0.63	0.03
370—270	10.3	0.65
(303.8, He II)	(5.4)	(0.35)
270—205	4.5	0.36
205—153	4.6	0.49
153—100	0.4	0.06
120—80	0.3	0.066
80—40	0.33	0.108

The solar photon fluxes vary both over a long period (11 year solar cycle) as well as over short periods (27 days) during disturbances. Although the exact amplitude of these variations is not yet fully established, it appears to be for the EUV range of the order of 2. The 10.7 cm solar radio flux is usually considered to be an excellent indicator of solar activity; according to limited data the integrated flux in the range 1310—270Å changes by a factor of 1.4—1.5 for a change in $S_{10.7}$ by a factor of 2 [54].

The solar cycle variation for the range from 10 to 100 Å amounts to a factor of 7, the flux in this range is ∼0.9 erg cm^{-2} sec^{-1} at solar maximum. The X-ray intensity at 8Å in the absence of solar flares varies over the solar cycle by a factor of 300; the variation in the range from 2—8 Å from a completely quiet sun to a class 3 solar flare can amount to five orders of magnitude (10^5), as shown in Table 12 [52a].

---

* The fluxes listed here are reduced by a factor of about 2 to 3 compared to earlier measurements [53]. This has lead to some problems in the interpretation of planetary ionospheres (see Chapter IX).
** See p. 19 for definition.

Table 12. *Variations in the X-ray intensities with solar activity*

Condition of the Sun		Intensities in $\mathrm{erg\,cm^{-2}\,sec^{-1}}$		
		$2\,\overset{\circ}{\mathrm{A}}$	$4\,\overset{\circ}{\mathrm{A}}$	$6\,\overset{\circ}{\mathrm{A}}$
Solar minimum	Completely quiet	10^{-8}	10^{-7}	10^{-6}
	Quiet	10^{-7}	10^{-6}	10^{-5}
	Lightly disturbed	10^{-6}	10^{-5}	10^{-4}
Solar maximum	Disturbed	10^{-5}	10^{-4}	10^{-3}
	Special events	10^{-4}	10^{-3}	10^{-2}
Flares	Class 3 Flares	10^{-3}	10^{-2}	10^{-1}

Resonance and Fluorescene Scattering of Solar XUV Radiation

In addition to dissociation and ionization, the interaction of solar radiation with a planetary atmosphere leads to emissions which are known as *day airglow* or *dayglow* [56, 57, 58]. Solar emission lines corresponding to resonance lines of atmospheric constituents lead to *resonance scattering*, i. e., the absorption of a solar photon leads to reradiation at the same wavelength. *Flourescence scattering* represents the case where the excited atom or molecule emits a photon at a wavelength longer than that of the absorbed photon. Airglow observations provide a sensitive technique for the detection of certain atmospheric constituents even though they may be extremely minor species of a planetary atmosphere [59, 60, 61]. Constituents of planetary atmospheres which may be detected by resonance and flourescence scattering observations are shown in the following Tables 13 and 14. In addition

Table 13. *Resonance reradiation*

$\mathrm{He^+}$	$304\,\overset{\circ}{\mathrm{A}}$
$\mathrm{Ne^+}$	$461\,\overset{\circ}{\mathrm{A}}$
He	$584\,\overset{\circ}{\mathrm{A}}$
$\mathrm{A^+}$	$724\,\overset{\circ}{\mathrm{A}}$
A	$1048\ (866)\,\overset{\circ}{\mathrm{A}}$
$\mathrm{N(N_2)}$	$1200\,\overset{\circ}{\mathrm{A}}\star\ (1135\,\overset{\circ}{\mathrm{A}})$
$\mathrm{H(H_2)}$	$1216\,\overset{\circ}{\mathrm{A}}\star\star$
O	$1304\ (1302,\ 1306)\,\overset{\circ}{\mathrm{A}}$
C	$1657\,\overset{\circ}{\mathrm{A}},\ 1561\,\overset{\circ}{\mathrm{A}}$

$\star$ also dissoc. excitation of $\mathrm{N_2}$
$\star\star$ also dissoc. excitation of $\mathrm{H_2}$

Table 14. *Fluorescence*

U.V.	(Bands)	X-rays	(K_a emissions)
CO	$1510\,\text{Å}$	C	$44.54\,\text{Å}$
NO	$2149\,\text{Å}$	N	$31.56\,\text{Å}$
N_2	$3914\,\text{Å}$	O	$23.57\,\text{Å}$
CO_2^+	$2882\,\text{Å}$, $3509\,\text{Å}$	Ne	$14.59\,\text{Å}$
CO^+	$2190\,\text{Å}$, $4264\,\text{Å}$	A	$4.19\,\text{Å}$

to their identification, the concentration of the scattering constituents can be determined from airglow observations [60].

The *column emission rate* $\mathscr{I}$ in photons $\text{cm}^{-2}\,\text{sec}^{-1}$ of an airglow layer is related to the column density of atoms or molecules through the emission rate factor g according to

$$\mathscr{I} = g \int_0^\infty n(R)\,ds. \tag{2-1}$$

The emission rate factor depends on the differential solar photon flux outside the atmosphere $\hat{\Phi}_\infty$ in photons $\text{cm}^{-2}\,\text{sec}^{-1}\,\text{Å}^{-1}$ and the oscillator strength of the transition, f:

$$g = \hat{\Phi}_\infty \lambda^2 \left(\frac{\pi e^2}{m c^2} \right) f.$$

(Note that $\Phi_\infty = \int_\lambda^{\lambda + \Delta\lambda} \hat{\Phi}_\infty\, d\lambda$, where $\Delta\lambda$ is the instrumental bandwidth; the numerical value for the constant in ( ) is 8.829×10^{-13} cm.) The column density in the line of sight (ds) can be related to the vertical column density by the Chapman function $\text{Ch}(\chi)$ (see p. 55)

$$\mathscr{I} = g\,\text{Ch}(\chi) \int_{R_s}^\infty n(R)\,dR = g\,\mathscr{N}_0\,\text{Ch}(\chi) \tag{2-2}$$

where R_s is the planetocentric distance of the observation point and $\mathscr{N}_0$ is the vertical column density.

For limb observations ($\chi = 90°$) outside★ a planetary exosphere whose density distribution can be approximated by a power law

---

★ In this case $R_s \gg R^*$, the minimum distance of the grazing ray, so that the *vertical* total content above R^* is related to the column density in the line of sight from $-\infty$ to $+\infty$, or in Equ. (2-4) by *twice* the Chapman function $\text{Ch}(\pi/2, R^*)$.

$n(R)=n(R^*)(R^*/R)^k$, the observed column emission rate is given by

$$\mathscr{I} = \mathscr{g}n(R^*)R^* \frac{2(\pi)^{\frac{1}{2}}}{k-1} \frac{\Gamma_\infty\left(\dfrac{(k+1)}{2}\right)}{\Gamma_\infty\left(\dfrac{k}{2}\right)} \tag{2-3}$$

where Γ_∞ is the complete gamma function.

The observed column emission rate for limb observations $(\chi=90°)$ of a constituent which follows an exponential altitude distribution $n(R)=n(R^*)\exp((R-R^*)/H)$ is given by

$$\mathscr{I} = \mathscr{g}\,n(R^*)\,(2\pi H R^*)^{\frac{1}{2}} \tag{2-4}$$

noting that $\mathrm{Ch}(\chi=90°, R^*)=((\pi/2)\,(R^*/H))^{\frac{1}{2}}$ (see also Chapter VIII).

From such observations, the density distribution $n(R)$ can be derived. It should be noted that the airglow emission rates are generally quoted in units of rayleighs; 1 rayleigh $(R)=10^6$ photons cm^{-2} sec^{-1}.

Production of Ionization by Solar Photons

In the photoionization process the production of an ion pair (ion + electron) occurs as the result of the action of a photon whose energy $\hbar v$ is equal to or greater than the ionization potential of the constituent X, according to

$$X+\hbar v(\geq \mathrm{IP}) \rightarrow X^+ +e.$$

For the general discussion of the production of electrons and ions by photoionization of neutral atmospheric constituents we will make some simplifying assumptions. The more general case can easily be obtained from this starting point by summation over all wavelengths, etc.

The following derivations are based on the assumptions of 1) monochromatic radiation, i. e., a small wavelength interval over which the absorption and ionization cross sections can be taken as constant; 2) an isothermal atmosphere $(H=\mathrm{const})$ and a horizontally stratified atmosphere (i. e., neglecting the curvature of the planetary surface). The ionizing radiation is assumed to be incident under an angle χ with the vertical (the solar zenith angle), having a flux outside the atmosphere Φ_∞. The ion production rate depends on the number density of the ionizable constituent n_j, its ionization cross section σ_i for the particular wavelength interval and the local photon flux Φ; according to

$$q = \sigma_i n_j \Phi. \tag{2-5}$$

Along a pathlength ds, $n_j \sigma_a ds$ photons are absorbed, where σ_a is the absorption cross section which is usually larger than σ_i. Hence we can write

$$\frac{d\Phi}{dz} = n_j \sigma_a \Phi \sec \chi$$

where $ds = dz \sec \chi$, and obtain by integration

$$\Phi = \Phi_\infty \exp \left[-\sigma_a \sec \chi \int_z^\infty n_j(z) dz \right], \tag{2-6}$$

i.e., the ionizing photon flux decreases from its value outside the atmosphere as the result of absorption. The ion production rate can now be expressed as

$$q = \sigma_i n_{j0} \Phi_\infty \exp \left[-\frac{z}{H} - \sigma_a \sec \chi \int_z^\infty n \, dz \right] \tag{2-7}$$

or in terms of the optical depth $\tau = \sigma_a \sec \chi \int_z^\infty n \, dz = \sigma_a n(z) H \sec \chi$

$$q = \sigma_i n_j(z) \Phi_\infty e^{-\tau}. \tag{2-8}$$

The ion production rate has a maximum where $dq/dz = 0$, which also corresponds to the condition $\tau = 1$, for overhead sun. Accordingly, the altitude where the maximum of ion production occurs can be found from this condition.

Since

$$\tau = \sigma_a n_0 H \sec \chi \cdot \exp \left(-\frac{z^*}{H} \right) = 1$$

we have $\exp(z^*/H) = \sigma_a n_0 H \cdot \sec \chi$ where it should be noted that z^* is the *relative* height above the reference level $z = 0$, represented by an absolute altitude h_0, usually close to or above the turbopause.

Hence the absolute altitude of the ion production maximum, h^* is given by

$$h^* = H \ln (\sigma_a n_0 H \sec \chi) + h_0 = h_0^* + H \ln \sec \chi. \tag{2-9}$$

The maximum value of the ion production rate is given by

$$q_m = \frac{\sigma_i}{\sigma_a} \frac{\Phi_\infty}{eH} \cos \chi = q_0 \cos \chi \tag{2-10}$$

where the ratio $\sigma_i/\sigma_a = \eta_i$ is called the ionization efficiency. The condition of $\tau = 1$ for overhead sun ($\chi = 0$) is often used to indicate the penetration depth of ionizing radiation, i.e., where it decreases by $1/e$

or where for $H = \text{const}$, the ionization rate would reach an absolute maximum

$$q_0 = \frac{\eta_i \Phi_\infty}{e\, H}.$$

The altitude h_0^* is also called *altitude of unit optical depth*, i.e., $\sigma_a n(h_0^*) H = 1$. It should be noted that the height of maximum ion production depends on the optical depth and solar zenith angle (sec χ), but not on the photon flux, whereas the maximum value of the ion production rate depends on the photon flux outside the atmosphere, the scale height and cos χ.

The ion production rate as a function of altitude can be written using the above definitions

$$q = q_m \exp\left\{ 1 - \frac{h - h^*}{H} - \sec\chi \exp\left[-\frac{(h - h^*)}{H} \right] \right\}. \tag{2–11}$$

This is the well known Chapman ion production function [62] which is illustrated in Fig. 21 in normalized form

$$q = q_0 \cos\chi \exp\left\{ 1 - Z - \sec\chi \exp(-Z) \right\} \tag{2–12}$$

where $Z = (h - h_0^*)/H$.

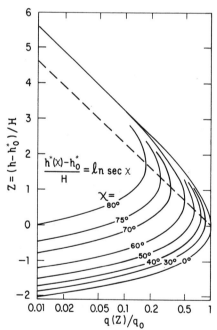

Fig. 21. Ion-pair production function, as function of solar zenith angle χ, normalized to the absolute ionization maximum q_0 for overhead sun ($\chi = 0$) occurring at an altitude h_0^*. The height parameter corresponds to $Z = (h - h_0^*)/H$

The optical depth and thus the penetration of ionizing radiation depends on the absorption cross section. In the EUV range, most major constituents of planetary atmospheres have absorption cross section of the order $\sigma_a \approx 10^{-17}$—$10^{-18}$ cm^2, whereas at X-ray wavelengths (see Table 15) the cross sections are much smaller [63,64].

Table 15

λ (Å)	σ_a (cm^2)	
1	$\sim 10^{-22}$	
3	$\sim 10^{-21}$	
5	$\sim 10^{-20}$	Between 20 Å and 50 Å the K ab-
15	$\sim 10^{-19}$	sorption limits occur, which cor-
50	$\sim 10^{-19}$	respond to an abrupt change in σ_a
100	$\sim 10^{-18}$	by about one order of magnitude.

The (total) absorption cross sections are the upper limit for the ionization cross sections; for atomic species $\sigma_a = \sigma_i$, i.e., the *ionization efficiency* $\eta_i = 1$, whereas for molecular species $\sigma_i \lesssim \sigma_a$, i.e., $\eta_i \lesssim 1$.

Figs. 22 to 26 show the absorption cross sections for important constituents of planetary atmospheres [63].

The ion production rate derived above under a number of simplifying assumptions can be extended to be more generally applicable. For an atmosphere with a constant scale height gradient $dH/dz = \beta$, the ionization maximum occurs where $\tau = 1 + \beta$ and the ion production rate is given by (cf. 1–12)

$$q = q_m (1 + \beta)\{1 - \zeta - e^{-\zeta} \sec \chi\}.$$

Near sunrise and sunset, i.e., $\chi > 75°$, the $\sec \chi$ term in the ion production function, which is a consequence of flat planet approximation,

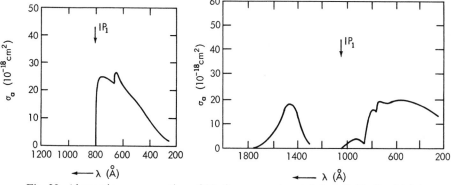

Fig. 22. Absorption cross section of N$_2$ from experimental data (left); O$_2$, (right). (After [63])

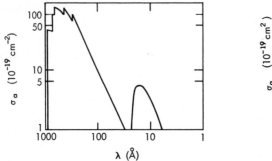
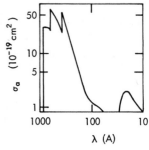

Fig. 23. Absorption cross section of O (left) and N (right) from theoretical calculations. (After [63])

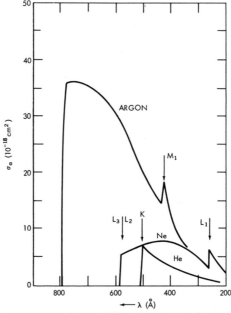

Fig. 24. Absorption cross sections of He, Ne and Ar from experimental data. (After [63])

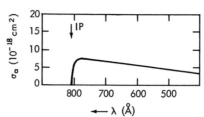

Fig. 25. Absorption cross section of H_2 based on experimental data. (After [63])

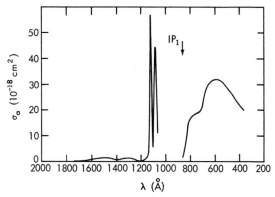

Fig. 26. Absorption cross section of CO_2 from experimental data. (After [63])

has to be replaced by the so called *Chapman function* $\mathrm{Ch}(\chi)$ [65]. This function is defined as the ratio of the total content of the atmosphere in the line of sight (s) to the sun to the vertical column content,

$$\mathrm{Ch}(\chi) = \frac{\int\limits_{s}^{\infty} n(s)\,ds}{\int\limits_{h}^{\infty} n(h)\,dh}.$$

A number of analytical approximations to the Chapman function have been developed, including the extension to a constant scale height gradient β. For an isothermal atmosphere a useful approximation of the Chapman function is given in terms of the tabulated error function [65]

$$\mathrm{Ch}\left(x, \chi \le \frac{\pi}{2}\right) \cong \left(\frac{\pi x}{2}\right)^{\frac{1}{2}} \left\{1 - \mathrm{erf}\left(x^{\frac{1}{2}}\cos\frac{\chi}{2}\right)\right\} \exp\left(x\cos^2\left(\frac{\chi}{2}\right)\right)$$

$$\mathrm{Ch}\left(x, \chi \ge \frac{\pi}{2}\right) \cong \left(\frac{\pi x}{2}\sin\chi\right)^{\frac{1}{2}} \left\{1 + \mathrm{erf}\left(-\mathrm{ctn}\,\chi\left(\frac{x\sin\chi}{2}\right)^{\frac{1}{2}}\right)\right\}$$

$$\cdot\left(1 + \frac{3}{8x\sin\chi}\right) \tag{2-13}$$

where $x = R/H = (R_0 + h)/H$ with H the scale height and R_0 the planetary radius. The Chapman function for different values of x and its comparison with $\sec\chi$ is shown in Fig. 27.

For $\chi = 90°$, $\mathrm{Ch}(x, \pi/2) = (x\pi/2)^{\frac{1}{2}}$, representing the ratio of the total content in the line of sight to the vertical content of an atmospheric constituent, a quantity of interest in occultation experiments.

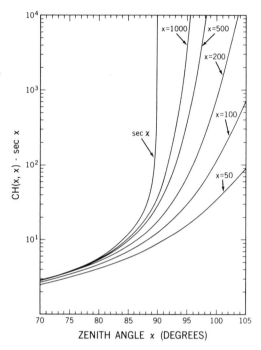

Fig. 27. Chapman function $Ch(x, \chi)$ for parametric values of $x = R/H$, compared with $\sec \chi$

At high enough altitudes where the optical depth is small the exponential factor in (2–8) can be neglected and the ion production function in this "low attenuation region" assumes the simple form

$$q = J(\sigma_i, \lambda) n_j(z) \qquad (2\text{–}14)$$

where $J = \langle \sigma_i \Phi_\infty \rangle$ is the *photoionization rate coefficient* (or frequency of ionization v_{ion}, $\sec^{-1}$). The ionization coefficient is based on the averaged ionization cross sections and photon fluxes over all pertinent wavelengths. For the light atmospheric constituents H and He, which will make up the main constituents of planetary exospheres, the ionization coefficients are listed in the following Table 16.

Table 16

J $(\sec^{-1})$	Venus	Earth	Mars
Hydrogen (H)	3×10^{-7}	1.5×10^{-7}	6×10^{-8}
Helium (He)	10^{-7}	6×10^{-8}	2.5×10^{-8}

Since many wavelengths will contribute to the ionization of a particular constituent n_j, and will also be absorbed by other atmospheric constituents, the ion production rate in this generalized form can be expressed

$$q = \sum_{j,\lambda} n_j \sigma_{ij} \Phi_\infty(\lambda) \exp\left(-\sum_l \sigma_a \int_z^\infty n_l \sec\chi\, dz\right). \qquad (2\text{–}15)$$

In the photoionization process, photoelectrons are produced whose energy $E_{pe} \cancel{h}\nu - \mathrm{IP}$, may be large enough, especially in the X-ray range, to lead to additional ("secondary") ionizations. In the X-ray range, the ion production is also computed by using the photon energy flux I_∞ divided by 35 eV, the average energy expended in the formation of an ion pair for air or CO_2. It should be noted that the calculation of secondary ionization depends on the initial photo-electron distribution and their collision processes with atmospheric particles. Photoelectrons with energies $\gtrsim 10\,\mathrm{eV}$ are also capable of exciting UV emissions (airglow) from the major constituents of planetary atmospheres [66]. Some planetary dayglow emissions excited by photoelectron impact are shown in the following Table 17.

Table 17. *Airglow emissions*

N_2	Second positive band	$3371\,\text{Å}$
N_2	Lyman-Birge-Hopfield band	$1354\,\text{Å}$
CO	Cameron band (a-X)	$1993\,\text{Å}$
CO	Fourth positive band (A-X)	$1478\,\text{Å}$
CO_2 (O; C)	(dissociative excitation)	$1356; 1561, 1657\,\text{Å}$

II.2. Corpuscular Radiation

Galactic cosmic rays, solar cosmic rays, solar wind protons and lower energy electrons and protons populating a planetary magnetosphere are responsible for the formation of electron-ion pairs at the lower level of planetary ionospheres [67]. Because of their high energy, cosmic rays penetrate deepest into the planetary ionosphere.

Galactic cosmic rays with energies $E > 10^9\,\mathrm{eV}$ are responsible for the formation of the lowermost ionospheric layer (sometimes called the cosmic ray or C layer) actually representing the lower part of the D layer.

Relativistic solar cosmic rays (protons) produced during highly disturbed solar conditions, which are a rather rare occurence (once in several years) having energies $E > 10\,\mathrm{BeV}$ cause a substantial enhancement in the ionization of the D layer [68].

Subrelativistic solar cosmic ray protons with energies $10\,\text{MeV} \lesssim E$ $\lesssim 100\,\text{MeV}$ produce strong ionization effects at polar caps which are accessible on a magnetic planet, leading to *polar cap absorption (PCA)* events in the terrestrial ionosphere. These events are generally observed at periods of maximum solar activity, but are almost non-existent during solar minimum [69]. The general formula for ion-pair production by charged particles (corpuscular radiation) can be written [70]

$$q(h) = \frac{1}{W} \int_E \int_\Omega \frac{dE}{dx} \cdot j(E)\,dE\,d\Omega \qquad (2\text{–}16)$$

where $W \simeq 35\,\text{eV}$ is the average energy required for the formation of an ion pair in air or CO_2, $dE/dx = \sec\chi\,dE/dh$, represents the energy loss in an inelastic collision process (ionization loss), $j(E) = K E^{-\gamma}$ represents the differential energy spectrum* of the ionizing particles and Ω is the solid angle. In the presence of a planetary magnetic field, the penetration of charged particles is inhibited; for a dipolar field only the most energetic particles can penetrate to low latitudes, so that the particle ionization effects are normally restricted to high magnetic (polar) latitudes. The ability of a charged particle with momentum (mv) to penetrate a magnetic field of strength B can be expressed by the *rigidity* P which is defined by

$$P = \frac{mvc}{Ze}\,(\text{volt}) = B\,r_B\,(\text{gauss cm})$$

where Z is the charge number, e is the electronic charge, and r_B is the Larmor radius of a charged particle of momentum mv moving in a plane perpendicular to the magnetic field vector $\boldsymbol{B}$.

The ion production rate resulting from galactic cosmic rays can be written in simplified form [67]

$$q(\varphi) = \frac{q_0(\varphi)n}{n_0}\,(\text{cm}^{-3}\,\text{sec}^{-1}) \qquad (2\text{–}17)$$

where $q_0(\varphi)$ is the ionization rate at magnetic latitude φ for 1 atmosphere with the corresponding number density $n_0 = 2.6 \times 10^{19}\,\text{mol cm}^{-3}$. The ionization rate at geomagnetic latitudes $\varphi \geq 70°$ corresponds to that which would prevail if the galactic cosmic rays were unimpeded by a planetary magnetic field; the ionization rate at the geomagnetic equator is reduced by an order of magnitude relative to that at the pole. The polar ionization rate at solar maximum has been found to be $q_0 \simeq 200\,\text{cm}^{-3}\,\text{sec}^{-1}\,\text{atm}^{-1}$, whereas that at solar minimum is by

---

* $j(E) = dJ/dE$, where $J(>E_0) = \int_{E_0}^{\infty} j(E)\,dE$ is the integral (spectrum) flux for energies grater than E_0.

about a factor of 4 larger, while that at middle and low latitudes varies by about a factor of 2, ranging from 150 to 300 cm^{-3} sec^{-1} at $\varphi = 50°$ [68]. This inverse relationship of cosmic ray intensity with solar activity, also called the solar modulation of cosmic rays, is due to the interplanetary magnetic field carried by the solar wind [69]. The penetration depth $\tilde{h}$ of charged particles in a planetary atmosphere is expressed by the *range* $\mathscr{R}$(g cm^{-2}), according to

$$\mathscr{R} = \sec \chi \int_{\tilde{h}}^{\infty} \rho(h)\,dh \qquad (2\text{–}18)$$

where χ is the zenith angle of the incident particles, and ρ is the mass density.

The range $\mathscr{R}$ is related to the ionization loss or *stopping power* according to

$$\mathscr{R} = \int_{E(\tilde{h})}^{E} \left[\frac{1}{\rho} \cdot \frac{dE}{dh} \right]^{-1} dE. \qquad (2\text{–}19)$$

Range-energy relations are usually based on experimental data or theoretical expressions based on the Bethe formula [71]. Empirical formulas exist for air, and since the pertinent physical parameters for CO_2 are very close to those for air, they may be applied to the atmospheres of the terrestrial planets. The range of protons with energies $E < 500$ MeV is approximately given by [72]

$$\mathscr{R}_p(\text{g cm}^{-2}) \cong \frac{E^{1.78}}{420}$$

and for electrons of energy $E < 200$ keV by

$$\mathscr{R}_e(\text{g cm}^{-2}) \cong \frac{E^{1.96}}{0.75}$$

where E is the particle energy in MeV.

Since charged particle trajectories are not rectilinear due to changes in direction in the collision process, as well as due to a planetary magnetic field, accurate analysis of energy deposition has to take these effects into account. Such calculations have been performed for low energy electrons using a statistical (Monte Carlo) method [73].

The approximate penetration depths at vertical incidence of energetic electrons and protons in the atmospheres of Earth, Venus and Mars, based on range-energy relations are shown in Fig. 28. The most penetrating charged particles are galactic and solar cosmic rays which produce ionization at altitudes below 100 km while lower energy particles, such as auroral electrons (1—10 keV) deposit their energy at altitudes

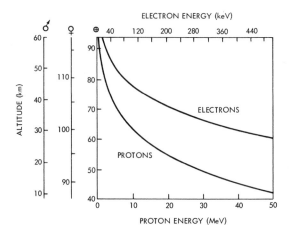

Fig. 28. Approximate altitudes of penetration of energetic particles in the atmospheres of Earth (⊕), Mars (♂) and Venus (♀)

100 km in the terrestrial atmosphere. Charged particles penetrate to
lower altitudes on Mars than on Earth, but not as deep into the atmosphere of Venus as on Earth. This is a consequence of the fact that in
the lower thermospheres of these planets the scale heights are not
greatly different, whereas the densities are highest for Venus and lowest
for Mars with Earth lying in between. For solar (cosmic ray) protons,
the use of an exponential rigidity spectrum of the form $J(>P)$
$=J_0 \exp(-P/P_0)$ with $P_0 < 100$ MV for polar or non-magnetic conditions and $P_0 > 150$ MV for geomagnetic latitudes $\varphi < 65°$, has been
found to give a good representation of their ionization rate [68]. Fig. 29
shows the ion-pair production functions over the polar cap for galactic
cosmic rays during solar maximum and minimum and a solar proton
event with a rigidity spectrum $J > 100$ MV $= 0.05$ cm^{-2} sec^{-1} ster^{-1} and
$P_0 = 50$ MV [68]. (Some solar proton events result in a peak ionization
rate 100 times larger than this.)

The solar (and galactic) cosmic ray ionization rate for Mars and
Venus [74] can be scaled from the terrestrial case according to

$$q(\tilde{h}_{M,V}) = \frac{\rho(\tilde{h}_{M,V})}{\rho(\tilde{h}_E)} q(\tilde{h}_E) \qquad (2\text{-}20)$$

where $q(\tilde{h}_E)$ is the ionization rate at the terrestrial poles, $\tilde{h}_E$ and $\tilde{h}_{M,V}$
are the atmospheric penetration altitudes resulting from the appropriate
range $\mathscr{R}(\text{g cm}^{-2})$ for the same cutoff energy for solar protons, and
$\rho(\tilde{h}_E)$ and $\rho(\tilde{h}_{V,M})$ are the respective mass densities.

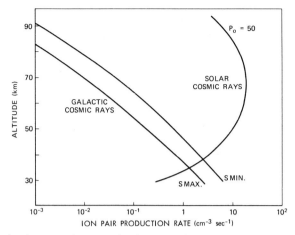

Fig. 29. Ionization rate due to galactic and solar cosmic rays in the terrestrial atmosphere over the polar cap. (After Webber [68])

For low energy electrons and protons, such as auroral particles, the ion production rate due to particles in the energy range E_{min} to E_{max} can be expressed by

$$q = \frac{\rho(h)}{0.032} \int_{E_{min}}^{E_{max}} L(E)j(E,h)\,dE \qquad (2-21)$$

where $\rho(h)$ is the atmospheric density (g cm^{-3}) at altitude h. The ionization loss for electrons can be approximated by [72]

$$L_e(E) = -\frac{dE}{d\mathcal{R}} = 0.38\,E^{-0.96}$$

and for protons by

$$L_p(E) = -\frac{dE}{d\mathcal{R}} = 236\,E^{-0.78}$$

where $\mathcal{R}$ is the range in g cm^{-2} and E is in MeV.

The number of particles of energy E arriving at altitude h, i.e., the differential energy spectrum $j(E,h)$ in cm^{-2} sec^{-1} ster^{-1} keV^{-1} has to be obtained by integrating over both pitch (α) and solid (Ω) angle if a magnetic field is present, according to

$$j(E,h) = \int_0^\Omega \int_0^{\frac{\pi}{2}} j(E)\exp\left[-\frac{x}{\lambda(E)\cos\alpha}\right]\sin\alpha\,d\alpha\,d\Omega \qquad (2-22)$$

where x is the range at height h and $\lambda(E) = 3.15 \times 10^{-7} E^{2.2}$ is the attenuation mean free path.

Another possible ionization source for planetary ionospheres is the solar wind. Although most of the solar wind flows around the planetary obstacle (cf. Chapter VI), a small fraction of the solar wind protons may leak through the magnetopause or ionopause, especially in the tail region.

Ionization by the solar wind is due to two processes: 1) direct impact ionization of atmospheric constituents by solar wind protons having an energy $E \sim 1\,\mathrm{keV}$, and 2) ionization by *hot* hydrogen generated by charge exchange (cf. Chapter IV) between solar wind protons and atmospheric constituents. The ionization rate due to the solar wind can be expressed by

$$q = \sigma_i' n_j \Phi_{sw}' + \sigma_i'' n_j \Phi_{H*}'$$

where the σ_i are the ionization cross sections ($\sim 10^{-16}\,\mathrm{cm}^2$), Φ_{sw}' is the solar wind proton *leakage* flux, i.e., a small fraction ($\lesssim 10\%$) of the solar wind flux outside the planetary magnetopause or ionopause Φ_{sw}, and Φ_{H*}' is the effective ionizing flux of hot hydrogen. The latter is governed by a continuity equation of the form

$$\frac{d\Phi_{H*}'}{ds} = -\frac{1}{2}\frac{d\Phi_{sw}'}{ds} - \frac{d\Phi_{H*}}{dE}\frac{dE}{ds}$$

where the factor $\frac{1}{2}$ denotes the downward part of an isotropic flux of hot hydrogen generated by charge exchange between solar wind protons and the neutral constituent n_j, according to $d\Phi_{sw}'/ds = -\sigma_{ex} n_j \Phi_{sw}'$, with $\sigma_{ex} \approx 10^{-15}\,\mathrm{cm}^2$; the second term on the right hand side represents the thermalization of the initially formed hot hydrogen beam along its path by scattering and ionization processes (cf. Chapter III).

Although it is difficult to assess the importance of solar wind ionization, since little is known about the solar wind penetration through magnetic "barriers", the solar wind appears to represent a significant source for the *nightside* of a planetary ionosphere like Venus (cf. Chapter IX).

II.3. Meteor Ionization

Meteoroids entering the upper atmosphere at high speeds ($v_0 \gtrsim v_\infty$) dissipate their energy and mass in the interaction with atmospheric molecules, producing the "visible" (either by eye or radar) meteor. The collisions with atmospheric molecules of the fast meteoric atoms and molecules, formed by ablation of the meteoroid, leads to excitation and ionization. Most of the meteors result from tiny pieces of compressed dust (thought to be of cometary origin) which are distributed in a continuous stream along the highly eccentric orbit around the sun. Thus, meteor streams intersect the planetary orbits out to the distance of Jupiter [75].

The ionization rate due to meteoroids of given initial mass, velocity and zenith angle distribution is given in the general form by [76]

$$q = \int_{\mu_{01}}^{\mu_{02}} \int_{v_{01}}^{v_{02}} \Phi\,\alpha(h,\mu_0,v_0,\chi)\sec\chi\,f_1(\mu_0)\,f_2(v_0)\,f_3(\chi)\,d\mu_0\,dv_0\,d\chi \qquad (2\text{--}23)$$

where Φ is the normalized total meteor flux, $\alpha(h,\mu_0,v_0,\chi)$ is the number of ion pairs produced per unit pathlength, i.e., the line density (cm^{-1}), $f_1(\mu_0)$, $f_2(v_0)$ and $f_3(\chi)$ are the distribution of meteoroid mass, velocity and of the meteor zenith angle, respectively. The integration limit μ_{01} refers to the minimum mass, represented by a boundary mass value between micrometeorites and meteoroids which suffer significant evaporation traveling through the atmosphere and which depends on the meteoroid mass density and velocity according to $\mu_{01} \sim \delta^{\frac{2}{3}}v_0^{-3}$; μ_{02} represents the mass of meteoroids which are heated uniformly throughout their volume while traveling through the atmosphere ($\mu_{02} \sim \delta_0^{-1}$). The meteoroid mass distribution is usually taken as $f_1(\mu_0) = 1/\mu^s$ with $s = 2$. The integration limits for velocity represent the minimum and maximum velocities in a planetary atmosphere, if it is assumed that the meteor streams are in heliocentric orbits; v_{01} therefore corresponds to the appropriate planetary escape velocity, while v_{02} represents a velocity which is the vector sum of the heliocentric escape velocity $v_\odot$ at the planet and the orbital velocity of the planet. Table 18 lists these values for a number of planets.

Table 18. *Characteristic velocities* (km/s)

	v_{01}	$v_\odot$	v_{orb}	v_{02} (max)
Venus	10	50	35	85
Earth	11	42	30	72
Mars	5	34	24	58
Jupiter	60	19	13	32★

The number of ion pairs formed per unit path length *(line density)* for a given meteoroid mass, density and velocity is given by [75]

$$\alpha = C\left(\frac{\mu_0}{\delta}\right)^{\frac{2}{3}}\rho v_0^4 \qquad (2\text{--}24)$$

★ At Jupiter, all meteoroids in heliocentric orbit will have the velocity v_{01}, since $v_{02} < v_\infty$.

where C is a constant depending on shape factor, ionization efficiency and heat properties of the meteor.

The atmospheric density where the maximum of the line density α_m occurs is given by $\rho(h^*) \sim \alpha_m/v_0^{\frac{1}{3}}$. Using this relation, the maximum of meteor ion production for a given mass, density and velocity is found to be given by the condition [75]

$$\rho(h^*) \cdot H \cong C^{\frac{3}{2}} \mu_0^{\frac{1}{3}} \delta^{-\frac{1}{2}} v_0^{-2} \tag{2–25}$$

where H is the atmospheric scale height. This condition also defines the penetration depth of meteors (h^*).

Detailed calculations of meteor ionization rates for the terrestrial atmosphere using equ. (2–23) with realistic distributions of meteor mass and velocity and an average meteor particle flux $\Phi(\mu_0 > 2 \times 10^{-4}\,\text{g})$ $= 10^{-4}\,\text{km}^{-2}\,\text{sec}^{-1}$ have been made [76]. The results of these calculations are illustrated in Fig. 30 and 31. Fig. 30 shows the ionization rates for given meteor velocities as a function of altitude; the meteor mass range contributing to the ionization according to the formulas for μ_{01} and μ_{02} is between 10^{-2} g and 3×10^{-13} g. Fig. 31 shows the total ionization rate from meteors, including that from micrometeorites, corresponding to a mass range from 10^{-13} g to 3×10^{-7} g and a velocity range from 15 km/sec to 70 km/sec.

The height of the meteor ionization maximum, i.e., the penetration depth of meteors in other planetary atmospheres can be scaled from the terrestrial results by virtue of the condition $(\rho_m H_m)_\text{E} = (\rho_m H_m)_\text{P}$. Thus,

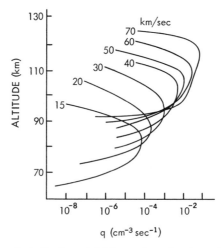

Fig. 30. Ionization rate in the terrestrial ionosphere due to meteors with different entry velocities. (After [76])

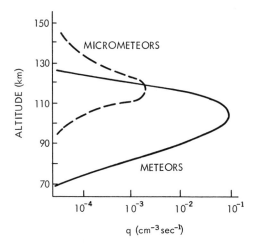

Fig. 31. Total ionization rate due to meteors and micrometeors in the terrestrial ionosphere. (After [76])

the penetration depth h_P^* for planet P is related to that for Earth h_E^* through the number density at the meteor ionization maximum and the ratios of the acceleration of gravity and atmospheric temperature at the appropriate level, according to

$$n(h_P^*) = n(h_E^*)\left(\frac{g_P}{g_E}\right)\left(\frac{T_E}{T_P}\right).$$

or the pressures according to

$$p(h_P^*) = p(h_E^*)\left(\frac{g_P}{g_E}\right).$$

The approximate altitudes of maximum meteor ionization for the terrestrial planets, where the relatively small ionization rates $(q_m \cong 10^{-2}$ to 10^{-1} cm^{-3} sec$^{-1})$ may contribute to the formation of an E-layer, are given in Table 19.

Table 19

Planet	h^* (km)
Venus	~ 115
Earth	~ 105
Mars	~ 90

The meteor flux due to cometary meteoroids, (and thus the corresponding meteor ionization rate) in the vicinity of a planet P can be
scaled according to $\Phi_P \propto D_P^{-1.5}$, where D_P is the planetary distance
from the sun in AU. In addition, the meteor flux near a planet is
enhanced due to gravitational effects by a factor $G = 1 + v_\infty^2(P)/v_0^2$.

Chapter III

Thermal Structure of Planetary Ionospheres

III.1. Departures from Thermal Equilibrium

Electrons released in the photoionization process due to the absorption of solar XUV radiation in planetary atmospheres may have initially kinetic energies of one to 100 eV. The energy E of photoelectrons resulting from photons of energy $\hbar\nu$ interacting with an atmospheric constituent whose ionization potential is IP is given by

$$E = \hbar\nu - \text{IP}.$$

These photoelectrons loose their excess energy by both inelastic and elastic collisions [77]. As long as the energy of the primary photoelectron exceeds IP, secondary ionization may occur, while below IP energy loss will occur by excitation of atmospheric constituents. The rate of energy loss per unit pathlength due to an *inelastic* collision with a neutral constituent whose density is n_j, can be expressed by

$$\frac{1}{n_j}\frac{dE}{dx} = E'\sigma_j \tag{3-1}$$

where E' is the excitation energy and σ_j is the appropriate cross section.

For the principal atmospheric constituents of the terrestrial planets, photoelectron energies $E > 20$ eV lead to optical excitation. For photoelectron energies $5 < E < 20$ eV, electronic excitation becomes the predominant energy loss process. Below 5 eV vibrational and at the lowest energies, rotational excitation of molecular species leads to the photoelectron energy loss by inelastic collisions with neutral particles. However, at energies below 2 eV, elastic Coulomb collisions are most important for the thermalization of photoelectrons. The rate of energy loss per unit path length for *elastic* collisions is given by

$$\frac{1}{N}\frac{dE}{dx} = -\frac{2m_e m}{(m_e + m)^2}\sigma_D E \tag{3-2}$$

where N is the number density and m the mass of the collision partner and σ_D is the collision (momentum transfer or diffusion) cross section

[63]. Due to the mass dependence, photoelectrons will lose their energy in elastic collisions primarily to the ambient electrons, rather than to the ions, leading to a preferential heating of the electrons gas, and an electron temperature greater than that of the heavy particles (ions and neutrals).

This can also be seen by considering the time constant for Maxwellianization of electrons or ions by self-interaction, according to

$$\tau_{jj} \propto m^{\frac{1}{2}} \frac{E^{\frac{3}{2}}}{N}. \tag{3-3}$$

Numerically, the self-collision time is given by [78]

$$\tau_{jj} = \frac{11.74 \, \hat{m}_j^{\frac{1}{2}} \, T^{\frac{3}{2}}}{N_j \, Z^4 \ln \Lambda}.$$

Where $\hat{m}_j$ is the mass in AMU, and Z the charge of the particles whose density is N and temperature T.

Since $\hat{m}_e = \frac{1}{1836} \hat{m}_p$, the self-collision time for electrons is by a factor $\frac{1}{43}$ less than that for protons; $\ln \Lambda$ is a parameter called Coulomb logarithm depending on T and N, having numerical values between 10 and 20 for typical conditions in planetary ionospheres (see Chapter VII).

Equipartition between the hotter electrons and the cooler ions will then take place by electron-ion (Coulomb) collisions.

The equipartition time, τ_{ei} is defined according to

$$\frac{dT_e}{dt} = -\frac{T_e - T_i}{\tau_{ei}} \tag{3-4}$$

and is related to the electron self-collision time τ_{ee} by

$$\tau_{ei} \simeq \frac{1}{2} \frac{\hat{m}_i}{\hat{m}_e} \tau_{ee} \simeq \frac{5.7 \, \hat{m}_e \hat{m}_i}{N \ln \Lambda} \left(\frac{T_e}{\hat{m}_e} + \frac{T_i}{\hat{m}_i} \right)^{\frac{3}{2}}.$$

Since $\tau_{ei} \gg \tau_{ee}$, the electron temperature T_e can exceed the ion temperature T_i, in altitude regions where the main cooling process for electrons is that of electron-ion collisions, as long as the electron gas is selectively heated.

The detailed thermal balance of a planetary ionosphere and the individual electron and ion temperature depend on the sources and sinks, both local and non-local, of the thermal energy of the ionospheric particles [79, 80, 81, 82]. These topics will be discussed in the following sections.

III.2. Electron and Ion Temperatures

The temperature of the electrons and ions of a planetary ionosphere is governed by heat balance equations of the form

$$\frac{3}{2} N \ell \frac{\partial T_{e,i}}{\partial t} + \sin^2 I \frac{\partial}{\partial z} \left(K_{e,i} \frac{\partial T_{e,i}}{\partial z} \right) = Q_{e,i} - L_{e,i} \qquad (3-5)$$

as long as the energy of any bulk motion is small compared to that of the random thermal energy. The subscript e and i refers to the electron and ion properties, respectively. K refers to the thermal conductivities and Q and L refer to the rates of heat production and loss. If the ionosphere is pervaded by a magnetic field, the heat conduction term is strongly controlled by the direction of this field; the $\sin^2 I$ factor, where I is the inclination (dip angle) of the magnetic field, implies that heat conduction is important primarily along magnetic flux tubes, but is strongly inhibited across the field (cf. Chapter VII).

Electron Heating

As stated in the previous section, the initial source of energy for heating the charged particles in the ionosphere derives from the photoionization process in which energetic photoelectrons are created which eventually thermalize. This leads to an enhanced electron temperature; the difference between electron and ion (and neutral) temperatures, i. e., the absence of thermal equilibrium in a planetary ionosphere is controlled by the detailed balance of sources and sinks of thermal energy for both the electrons and ions. The initial photoelectron energy spectrum, which depends on the characteristics of the ionizing solar XUV radiation and the composition and density distribution of the neutral atmosphere and appropriate cross sections represents the starting point for the ionospheric heat input. The local heating rate Q is related to the local ionization rate q by

$$Q_e = \varepsilon q \qquad (3-6)$$

where ε is the energy per ion pair formation which is transferred locally to the electron gas in form of heat.* From the previous definition of the ion production rate (Chapter II) it can be seen that the local heating rate has an altitude dependence which is largely controlled by that of

---

* $\varepsilon q = \int\limits_{E_{th}}^{\infty} q(E) dE / [1 + (dE/dx)_n / (dE/dx)_e]$, where E is the initial photoelectron energy, E_{th} is the thermal energy of the electron gas and the subscripts n and e refer to energy loss rate to neutrals and electrons, respectively.

the ion production rate, and is thus given by

$$Q(z) = \varepsilon(z) n_0 \, \sigma_i \, \Phi_\infty \exp\left[-\frac{z}{H}\right].$$

At altitudes where inelastic collisions of electrons with neutrals are important, the heating efficiency is roughly inversely proportional to the neutral gas density, while at higher altitudes ε is strongly dependent on the electron, ion and neutral density. At these higher altitudes, specifically, where the electron mean free path $\lambda_{e,n}$ becomes comparable to the scale height of the neutral gas responsible for scattering, the photoelectrons can escape, i. e., travel significant distances before becoming thermalized [80]. This photoelectron escape flux is therefore responsible for non-local heating of the electron gas. The escape level for photoelectrons, in analogy to the definition of the neutral exobase, is defined by the condition $\lambda_{e,n} = H$ or $n = (\sigma_{sc} H)^{-1}$ where σ_{sc} is the scattering cross sections for electrons in the neutral gas ($\sigma_{sc} \cong 10^{-15}$ cm^2). The magnitude of the photoelectron escape flux can be estimated from simple scattering considerations [82]. The continuity equation for the photoelectron flux ϕ can be expressed by

$$\frac{d\phi}{ds} = \tfrac{1}{2} q - L(s) \tag{3–7}$$

where q is again the production rate of photoelectrons and L represents the loss due to electron-neutral particle scattering along the path s.

The factor $\tfrac{1}{2}$ is based on the assumption of an initial isotropic photoelectron distribution, so that one half of the electrons produced at a given level where escape can take place are directed upward. If there is a planetary magnetic field pervading the ionosphere, the path of the photoelectrons is constrained by field lines and the scattering also depends on the local pitch angle α. For a probability of 0.5 for an inelastic collision to remove a photoelectron from the flux, the loss rate can be expressed by

$$L = \tfrac{1}{2} n \, \sigma_{sc} \, \phi \, \sec \alpha.$$

Integration of (3–7) gives for the unscattered photoelectron escape flux for the limit $(z \to \infty)$

$$\phi_\infty = q \, \lambda_{e,n} \cos \alpha$$

or (for $\alpha = 0$),

$$\phi_\infty = \frac{\sigma_i}{\sigma_{sc}} \Phi_\infty = \frac{J}{\sigma_{sc}}. \tag{3–8}$$

With ionization rate coefficients $J \simeq 10^{-7}\,\text{sec}^{-1}$ and scattering cross sections $\sigma_{sc} \simeq 10^{-15}\,\text{cm}^2$, photoelectron escape fluxes of the order of $\phi_\infty \simeq 10^8\,\text{cm}^{-2}\,\text{sec}^{-1}$ are obtained, and photoelectron energy fluxes of the order of $10^9\,\text{eV}\,\text{cm}^{-2}\,\text{sec}^{-1}$, assuming an average photoelectron energy of 10 eV. Such energy fluxes can lead to heating rates at high altitudes which are orders of magnitudes larger than those resulting from local ion production. Thus, escaping photoelectrons can represent an important source of heating in the upper portions of planetary ionospheres, and can also lead to heating of a nighttime ionosphere, if the magnetically conjugate point is already sunlit [83, 84].

The heating rate due to a flux of photoelectrons is given by

$$Q_{pe}(z) = \int\limits_0^\infty \left(\frac{d\phi}{dE}\right)\left(\frac{dE}{dz}\right)dE \qquad (3\text{--}9)$$

where dE/dz is the energy loss of photoelectrons due to elastic collisions with ambient electrons, leading to heating of the electron gas.

The energy loss of photoelectrons due to elastic collisions has usually been determined from the formula of Butler and Buckingham [cf. 77, 81]

$$\frac{dE}{dt} \equiv v\,\frac{dE}{dz} = -\frac{8\sqrt{\pi}}{m_e}\,\frac{Ne^4}{v_{th}}\,\mathscr{F}\left(\frac{v}{v_{th}}\right)\ln\Lambda \qquad (3\text{--}10)$$

where

$$\mathscr{F}(u) = \frac{2}{\sqrt{\pi}}\left(\int\limits_0^u e^{-x^2}dx - 2u\,e^{-u^2}\right),$$

v_{th} is the electron thermal speed and $\ln\Lambda$ is the Coulomb logarithm. For energies $E > 3\,\text{eV}$ this equation can be approximated by [82]

$$\frac{dE}{dt} \simeq -\frac{10^{-5}}{E^{\frac{1}{2}}}\,N_e\;(\text{eV}\,\text{sec}^{-1}). \qquad (3\text{--}11)$$

In the presence of a magnetic field the right hand side of (3–11) is modified by a factor $(\cos\alpha\,\sin I)^{-1}$.

The formula of Butler and Buckingham neglects the energy loss of photoelectrons due to the generation of plasma waves by the Čerenkov mechanism (see Chapter VII); the complete expression for the energy loss of fast photoelectrons to thermal electrons is given by [85, 86]

$$\frac{dE}{dt} = -\frac{\omega_N^2 e^2}{v}\left\{\mathscr{F}\left(\frac{v}{v_{th}}\right)\ln\frac{2\Lambda}{3\gamma} + \frac{v}{v_{th}}\,\mathscr{G}\left(\frac{v}{v_{th}}\right)\right\} \qquad (3\text{--}12)$$

where $\mathcal{G}(v/v_{th})$ represents the energy loss due to plasma wave generation and ω_N is the plasma frequency $\omega_N = (4\pi N e^2/m_e)^{\frac{1}{2}}$. For energies $E \gg \mathcal{k} T$, this formula can be approximated by

$$\frac{dE}{dt} = -\frac{\omega_N^2 e^2}{v} \cdot \begin{cases} \ln\left(\dfrac{mv^3}{\gamma e^2 \omega_N}\right) & \text{for } \mathcal{k} T \ll E < \dfrac{m e^4}{2\hbar^2} \\[3mm] \ln\left(\dfrac{mv^2}{\hbar \omega_N}\right) & \text{for } E > \dfrac{m e^4}{2\hbar^2} \end{cases} \qquad (3\text{--}13)$$

where $2\pi \hbar \equiv$ Planck's constant and $\gamma = 1.781$ is Euler's constant. Energy loss rates based on the Butler and Buckingham formula as applied by many authors, may be as much as 70% lower than those based on (3–13) [85, 87].

A useful analytic form of eqn. (3–13) is given by [88]

$$\frac{dE}{dt} = -\frac{2 \times 10^{-4} N^{0.97}}{E^{0.44}} \left(\frac{E - E_e}{E - 0.53 E_e}\right)^{2.36} \qquad (3\text{--}14)$$

where

$$E_e = 8.618 \times 10^{-5} T_e \,.$$

Photoelectron excitation of neutral constituents also produces *airglow emissions* of otherwise forbidden lines, e. g., the 1356/1358Å doublet of O and the 2972Å line of O; the oxygen green line (5577Å) and red line (6300Å) can also be produced by photoelectron excitation. Another example is the 10830Å line of metastable He, the Cameron $a-X$ bands of CO (1900—2500Å), the first negative bands of CO$^+$ (2100—2400Å) as well as the first negative system of N_2 (3914Å).

Electron Cooling

The thermal electrons can loose energy by elastic collisions with neutrals and ions, as well as by inelastic collisions with neutral species, leading to rotational, vibrational or electronic excitation of neutrals [77, 89, 89a].

The heat loss or cooling rate of the electrons can be expressed in general form by

$$L_e = \frac{d}{dt}\left(\frac{3}{2} N_e \mathcal{k} T_e\right) = \int_{E_0}^{\infty} v \frac{dE}{dx} f(E, T_e) dE \equiv \int_{E_0}^{\infty} \frac{dE}{dt} f(E, T_e) dE \qquad (3\text{--}15)$$

where dE/dx is the loss rate per unit path travelled for the appropriate collision process and $f(E, T_e)$ is the Maxwellian velocity distribution for the temperature T_e. The heat loss of electrons by elastic collisions with neutrals is given by

$$L_e(\text{elast.}, n_j) = -\frac{2 m_e}{m_j} N_e v_{en} \frac{3}{2} k (T_e - T_n) \qquad (3\text{-}16)$$

where v_{en} is the collision frequency between electrons and neutrals, which depends on the momentum transfer cross section σ_D acording to

$$v_{en} = \frac{4}{3} n_j \left(\frac{8 k T_e}{\pi m_e}\right)^{\frac{1}{2}} \sigma_D .$$

Because of the mass factor $2 m_e/m_j$, this particular cooling rate is rather small; energy loss of electrons occurs much more rapidly as the result of inelastic collisions. Electron cooling by rotational excitation of non-polar molecules such as N_2, O_2 and CO_2 can be expressed by

$$L_e(\text{rot.}) = -a_j N_e n_j T_e^{-\frac{1}{2}} (T_e - T_n)^\star \qquad (3\text{-}17)$$

where the constant a_j has values of the order 10^{-14} for the previously mentioned nonpolar molecular species. For polar molecules such as CO, the relation (3–17) does not apply and cooling rates have to be derived from a general expression for energy transfer by inelastic collisions [77, 89]. The electron heat loss by vibrational excitation of molecular species is determined by the relation

$$L_e(\text{vib.}) \cong N_e n_j \exp\left(-\frac{\Delta E}{k T_n}\right) \left[\exp\left(\frac{\Delta E}{k T_n} \cdot \frac{T_e - T_n}{T_e}\right) - 1\right] \qquad (3\text{-}18)$$

where ΔE is the energy difference of two vibrational states. For typical conditions in planetary ionospheres, vibrational cooling of the electron gas may become important when the electron temperature is high. Similarly, electron heat transfer due to electronic excitation of atmospheric constituents, becomes primarily important for very high electron temperatures.

A very important energy transfer process at electron temperatures normally found in planetary ionospheres $(T_e \sim 1000\,^\circ\text{K})$ is that due to the excitation of fine structure transitions in atomic oxygen (O). The electron cooling rate for fine structure excitation of the ground state of atomic oxygen can be expressed by [cf. 89]

$$L_{e,n}(\text{f. str. O}) = -3 \times 10^{-12} N_e n(\text{O}) \cdot T_n^{-1} (T_e - T_n) .^\star \qquad (3\text{-}19)$$

The electron cooling rate resulting from heat transfer to another species can also be expressed by (3–4)

$$\frac{d T_e}{d t} = -\frac{T_e - T_j}{\tau_{ej}}$$

$\star$ (eV cm^{-3} sec^{-1}).

where τ_{ej} is the equilibration time. At some altitude where the electron-ion density is large enough, the equilibration time between electrons and ions τ_{ei} (3–4) will be less than that between electrons and neutrals, and the electron cooling will occur primarily by elastic (Coulomb) collisions between electrons and ions. The electron cooling rate due to Coulomb collisions is given by

$$L_{e,i} = -7.7 \times 10^{-6} N_e N_i \left(\frac{T_e - T_i}{\hat{m}_i T_e^{\frac{3}{2}}} \right) (\text{eV cm}^{-3} \text{sec}^{-1}). \qquad (3\text{--}20)$$

The energy transfer rate has a critical value for $T_e = 3 T_i$ corresponding to [79, 80]

$$L_{e,i}^* = -3 \times 10^{-16} \frac{N_e N_i}{\hat{m}_i T_i^{\frac{1}{2}}} (\text{eV cm}^{-3} \text{sec}^{-1}). \qquad (3\text{--}21)$$

For a local heat input into the electron gas $Q_e > L_{ei}^*$, T_e would not be limited by energy transfer to the ions, and would rise ("runaway") until controlled by other loss processes or heat conduction.

Ion Heating and Cooling Processes

The electron cooling rate due to Coulomb collisions provides the principal heat input to the ions. Accordingly, we have for the ion heating rate

$$Q_i = 7.7 \times 10^{-6} N_e N_i \left(\frac{T_e - T_i}{\hat{m}_i T_e^{\frac{3}{2}}} \right) (\text{eV cm}^{-3} \text{sec}^{-1}). \qquad (3\text{--}22)$$

Cooling of the ions takes place by energy transfer to the neutrals; the rate of heat transfer between ions and neutrals due to the temperature difference of the two gases, as well as that due to the possible difference of the bulk velocities v_i, v_n, (frictional heating) can be expressed by [89]

$$L_{in} = 2 N_i \frac{m_i m_n}{(m_i + m_n)^2} v_{in} \left\{ -\frac{3}{2} k (T_i - T_n) + \frac{1}{2} m_n (v_i - v_n)^2 \right\} \qquad (3\text{--}23)$$

where v_{in} is the ion-neutral collision frequency

$$v_{in} \cong 2 \pi \left(\frac{\alpha e^2}{\mu_{in}} \right) n. \qquad (3\text{--}24)$$

α is the polarizability of the neutral constituent whose density is n and μ_{in} is the reduced mass. Ions in their parent gas, also experience cooling by resonance charge transfer. In this case v_{in} is modified by a factor $(T_i + T_n)^m$; for most atmospheric gases $0.3 \lesssim m \lesssim 0.4$ [90].

Heat Transport in the Ionosphere

In addition to local heat input and loss by energy transfer processes (collisions), heat can also be transported in the ionospheric plasma by conduction. The change in temperature is then given by the divergence of the heat flux, which is expressed by

$$\mathscr{F}_{e,i} = K_{e,i} \nabla T_{e,i}. \tag{3-25}$$

For a fully ionized plasma, the thermal conductivity is primarily due to the random motions of the electrons and is given by [82]

$$K_e = 7.7 \times 10^5 \, T_e^{\frac{5}{2}} \, (\text{eV cm}^{-1}{}^\circ\text{K}^{-1} \sec^{-1})\star. \tag{3-26}$$

Since planetary ionospheres are not fully ionized plasmas, particularly at lower altitudes where electron-neutral collisions occur, a more appropriate form for the electron conductivity is [82]

$$K_e' = \frac{K_e}{1 + K_e/K_{en}} \tag{3-27}$$

where

$$K_{en} = \frac{57 \, N_e \, T_e^{\frac{1}{2}}}{\sum_j n_j \sigma_D}$$

with σ_D the momentum transfer cross section for the neutral constituents. The thermal conductivity for the ion gas is much smaller due to the smaller thermal velocity of the ions. The ion conductivity is given by

$$K_i = 4.6 \times 10^4 \frac{T_i^{\frac{5}{2}}}{\hat{m}_i^{\frac{1}{2}}} \, (\text{eV cm}^{-1}{}^\circ\text{K}^{-1} \sec^{-1}) \tag{3-28}$$

where $\hat{m}_i$ is the ion mass in AMU.

When a magnetic field is present, the heat flow is parallel to the field lines; in any other direction the thermal conductivity is reduced by $\sin^2 I$. The parallel thermal conductivity $K_{\parallel}$ is essentially the electron conductivity K_e, whereas perpendicular to the field lines, the ion conductivity predominates. The perpendicular conductivity $K_{\perp}$ can be expressed by

$$K_{\perp} \sim K_{\perp i} = K_{\parallel} \frac{\nu_{ei}}{\Omega_B^2} \left(\frac{m_i}{m_e}\right)^{\frac{1}{2}} \approx 3 \times 10^{-16} \frac{N_i^2}{B^2 \, T_i^{\frac{1}{2}}} \, (\text{erg} \, {}^\circ\text{K}^{-1} \, \text{cm}^{-1} \, \sec^{-1}) \tag{3-29}$$

where Ω_B is the ion gyrofrequency, $\Omega_B = e B/m_i c$.

$\star$ Also expressed in $\text{erg} \, {}^\circ\text{K}^{-1} \, \text{cm}^{-1} \, \sec^{-1}$ ($1 \text{ eV} = 1.6 \times 10^{-12}$ erg).

The ratio $K_\perp/K_\parallel$ is extremely small. A more efficient heat transfer across field lines may occur when the plasma is turbulent. This may apply in cases where the solar wind interacts directly with the planetary ionosphere (e. g., Venus). The "turbulent" heat conductivity K_B should be related to the *Bohm diffusion* coefficient

$$D_B \approx \frac{c k}{16 e}\frac{T}{B} \cong 5.4 \times 10^2\,\frac{T}{B}\,(\text{cm}^2\,\text{sec}^{-1})$$

where B is the magnetic field strength in gauss; thus, K_B is given by

$$K_B \approx k N D_B \approx 10^{-13}\,N\,\frac{T}{B}\,(\text{erg}\,{}^\circ\text{K}^{-1}\,\text{cm}^{-1}\,\text{sec}^{-1}) \qquad (3\text{--}30)$$

which is much greater than $K_\perp$ [91].

Equilibrium Electron and Ion Temperatures

At altitudes where heat input into the electron gas is roughly balanced by the energy transfer to ions and where conduction can be neglected in first approximation (e. g., in the vicinity of the ionization peak), the electron temperature varies according to

$$T_e - T_i \propto \frac{Q}{N^2}\propto \frac{1}{N} \qquad (3\text{--}31)$$

since the electron heating rate $Q \propto N$, while the loss rate $L_e \propto N^2$, where $N = N_e = N_i$. Thus, the electron temperature will be somewhat lower at F region altitudes, when the electron density is high such as at solar maximum conditions (Fig. 32). The ion temperature under equilibrium conditions neglecting conduction can be expressed by [82]

$$T_i = \frac{T_n + a\,\dfrac{N}{n}\,T_e^{-\frac{1}{2}}}{1 + a\,\dfrac{N}{n}\,T_e^{-\frac{3}{2}}} \qquad (3\text{--}32)$$

where

$$a \simeq 10^7.$$

At low ionospheric altitudes where N/n is small, the ion temperature will be equal to the neutral gas temperature $(T_i = T_n)$, due to the efficient energy transfer between ions and neutrals; at high altitudes where N/n becomes large, the ion temperature will approach the electron temper-

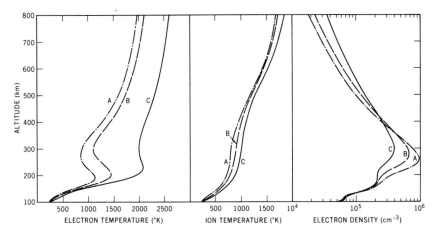

Fig. 32. Electron and ion temperatures in the terrestrial ionosphere for different conditions. Also shown are the corresponding electron density profiles. Curves A reflect a high peak electron density (e. g., solar maximum) and high atomic oxygen concentration at the lower boundary; curves C reflects a low peak electron density (e. g., solar minimum) and an atomic oxygen concentration reduced by 50 % while curves B reflects and intermediate condition. Similarly, curves C are also representative of a situation occurring during magnetic storms. (After Herman and Chandra)

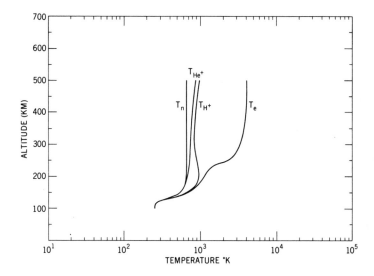

Fig. 33. Model temperature profiles of the neutrals (CO_2), the ions He^+ and H^+ and the electrons in the Venus ionosphere. (After [95])

ature $(T_i \rightarrow T_e)$ since Coulomb collisions between electrons and ions are the predominant energy transfer process.

At high altitudes in a planetary ionosphere, the electron temperature is largely controlled by heat conduction, since heat loss by energy transfer becomes negligible, while heat input occurs due to the presence of suprathermal particles (photoelectrons).

The heat flux

$$\mathscr{F} = \int_z^\infty Q(z)\,dz \tag{3-33}$$

can maintain a positive temperature gradient according to

$$\mathscr{F}_e = -K_e \frac{dT_e}{dz}. \tag{3-34}$$

Along a magnetic field line the maximum of the electron temperature occurs near the equator. A positive temperature gradient can also be maintained if heat is conducted into the ionosphere from an adjoining hot plasma, such as in the case of the solar wind interacting with a planetary ionosphere. The preceeding Figs. 32 and 33 show examples of steady-state electron and ion temperatures for the ionospheres of Earth and Venus, illustrating the role of the various processes discussed before in determining the thermal structure of a planetary ionosphere.

III.3. Heat Sources not Related to Photoionization

In addition to the principal heat source for the ionospheric plasma, viz. excess energy imparted to the photoelectrons in the photoionization process, other sources of energy input into the electron and ion gas exist.

Corpuscular Heating

A stream of fast electrons, as may exist in the magnetosphere, having an energy $E(\mathrm{eV})$ which interact with the thermal ambient electrons, can heat them at a rate

$$Q_e \cong 2 \times 10^{-12} \frac{N_e}{E} \Phi_e \, (\mathrm{eV \ cm^{-3} \ sec^{-1}}) \tag{3-35}$$

where Φ_e is the flux in $\mathrm{cm^{-2} \ sec^{-1}}$. This is essentially the same effect as heating of the ambient electrons by photoelectrons, except that the fast electrons considered here have their origin in processes other than photoionization (e. g., acceleration by electrostatic fields).

Joule Heating

In regions of a planetary ionosphere containing a magnetic field, where electric currents flow (e.g., the auroral and equatorial electrojet in the terrestrial ionosphere) conversion of electric current energy into thermal motion through charged particle collisions (Joule dissipation) can lead to heating of the charged particles and the neutrals [92, 93].

Joule heating is generally expressed by [92]

$$Q = j \cdot E \tag{3–36}$$

and in terms of conductivity by

$$Q = \sigma_1 E^2 = \frac{j}{\sigma_3} \tag{3–37}$$

where j is the current flowing in the ionospheric plasma as the result of the electric field E; σ_1 is the Pedersen and σ_3 the Cowling conductivity (cf. V.2).

The heating rate for electrons, for the case where the electric field E is perpendicular to the magnetic field can be expressed by [82, 94]

$$Q_e = \frac{E_\perp^2 c^2}{B^2} \left[N_e \sum_j v_{ej} + N_i \frac{v_{ei}}{1 + \sum \frac{\Omega_B}{v_{ij}}} \right] \tag{3–38}$$

where B is the magnetic field strength, v is the collision frequency with the subscripts e, i, j referring to the electrons, ions and neutrals, respectively, and Ω_B is the ion gyrofrequency, $\Omega_B = e B / m_i c$.

Similarly, the heat input into the ions is given by [82]

$$Q_i = \frac{E_\perp^2}{B^2} c^2 \sum N_i \cdot \frac{\sum_j \mu_{in} v_{ij}}{1 + \sum \frac{v_{ij}}{\Omega_B}} \tag{3–39}$$

where μ_{in} is the reduced mass $m_i m_j / (m_i + m_j)$.

In the absence of other heat sources for the electrons, the ion-heating due to (3–39) would lead to ion temperatures greater than T_e.

Solar Wind Heating

The hot solar wind plasma in the transition region between the bow shock and the "ionopause" (see Chapter VI), particularly for non-magnetic planets, can represent an additional heat source for the planetary ionosphere. The interaction between the solar wind and the planetary obstacle leads to the generation of hydromagnetic and acoustic energy.

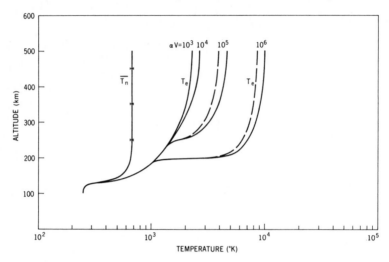

Fig. 34. Effect of solar wind heating on the Venus ionosphere; α is the heating efficiency and V the energy transport velocity (hydromagnetic or acoustic). The solid lines refer to a scale length for the energy dissipation $L = 300$ km, whereas the dashed lines represent $L = 100$ km. For $\alpha V = 10^4$, the solar wind heating is about 50% of the total heat input into the Venus topside ionosphere. (After [95])

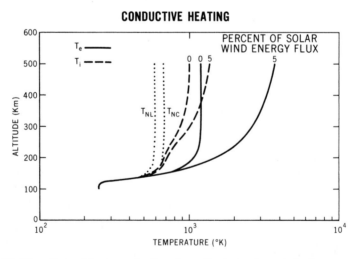

Fig. 35. The effect of heat conduction from the hot solar wind plasma on the electron and ion temperatures of the Venus ionosphere, when 5% of the solar wind energy flux is conducted into the topside ionosphere. T_{NL} and T_{NC} refer to the neutral gas temperatures for the light constituents and the main (CO_2) gas, respectively. (Courtesy of R. E. Hartle)

This energy can be transported in form of waves of the appropriate modes into the planetary ionosphere. If we assume that a fraction α of the solar wind energy, represented by its pressure p_{sw} (or equivalently its energy density), is deposited with a scale length L over an altitude range from the ionopause level z_u downward and a velocity V of the waves, (hydromagnetic, ion-acoustic, etc.) whose damping leads to the heating, the solar wind heat input can be expressed by [95]

$$Q_{sw} = \left(\frac{\alpha V p_{sw}}{L}\right) \exp\left\{\frac{(z - z_u)}{L}\right\}. \qquad (3\text{-}40)$$

If the waves are hydromagnetic in nature, the heating may occur for both electrons and ions, while acoustic modes will heat the ions preferentially (see Fig. 34).

Another source of solar wind heating is heat conduction from the hot solar wind plasma outside the ionopause into the topside planetary ionosphere. In this case a turbulent heat conduction coefficient may have to be invoked (3-30), since an even small horizontal magnetic field in the ionopause region $(B > 10^{-1}\gamma)$* may otherwise inhibit heat conduction. Some model calculations of the effects of solar wind heat conduction heating of the Venus ionosphere are illustrated in Fig. 35.

---

* $1\gamma = 10^{-5}$ gauss.

Chapter IV

Chemical Processes

IV.1. General Comments

Chemical reactions provide the only true sink of ionization by *recombination* of ion pairs. In addition to this loss process, *charge exchange* reactions act as sources or sinks for particular ion species without perturbing the overall ionization balance [96, 97].

The rate of change of ionized constituents resulting from a *recombination* process can be expressed by

$$\frac{\partial N_{ei}}{\partial t} = -\alpha N_i N_e = -\alpha N^2 \quad \text{(for } N_e = N_i \equiv N) \tag{4-1}$$

where α is the recombination coefficient, which is temperature-dependent. Loss of ion pairs by recombination is therefore represented by a square law. Solution of (4–1) gives

$$\frac{1}{N} = \frac{1}{N_0} + \alpha t. \tag{4-2}$$

The time constant* for recombination can be expressed by

$$\tau_{C\alpha} = \frac{1}{\alpha N}. \tag{4-3}$$

The rate of change of an ion species due to a *charge exchange* process is given by

$$\frac{\partial N_i}{\partial t} = -kn(M)N_i \tag{4-4}$$

where M refers to the neutral species involved in the charge exchange process; k is the rate coefficient, which can be expressed by

$$k = \langle \sigma_r v_r \rangle \tag{4-5}$$

---

* $\tau \equiv ((1/N)(\partial N/\partial t))^{-1}$.

with σ_r the cross section for the reaction and v_r the relative speed between the reacting particles, averaged over the velocity distribution function. The product $kn(M) = \beta$ is often called the *loss coefficient*, which is altitude-dependent, since it depends on the density of the constituent M.

Solution of (4–4) yields

$$N_i = N_{io} e^{-\beta t}. \tag{4–6}$$

Charge transfer reactions correspond to a linear loss law

$$\frac{\partial N}{\partial t} = -\beta N. \tag{4–7}$$

The time constant for charge exchange is given by

$$\tau_{C\beta} = \frac{1}{\beta} = \frac{1}{kn(M)}. \tag{4–8}$$

The controlling chemical process will be determined by the shortest time constant.

Recombination and charge exchange are representative of *two-body (binary) reactions*, i.e., involving two collision partners, for which the rate equation for the concentration of constituent $[X]$ can be expressed in general form as

$$\frac{d[X]}{dt} = -k[X][Y] \tag{4–9}$$

where the rate coefficient k has the dimension $cm^3 sec^{-1}$. At altitudes where the atmospheric pressure is relatively high ($p \geq 10^{-2}$ Torr), i.e., in the D-region of planetary ionospheres, *three-body reactions* can become important. In this case the rate equation can be expressed in general form by

$$\frac{d[X]}{dt} = -k_3[X][Y][Z] \tag{4–10}$$

where the three-body reaction coefficient has the dimension $cm^6 sec^{-1}$. Similarly, the rate of a one-body (spontaneous) process, such as photo-detachment, can be expressed by

$$\frac{d[X]}{dt} = -v[X] \tag{4–11}$$

where v is the frequency (sec^{-1}) of the process.

However, two- and three-body processes can also be expressed in the form of a linear loss law appropriate to a one-body process; in that case the frequency of the process, which then will *not* be a constant, is given by $v = k[Y]$ or $k_3[Y][Z]$.

IV.2. Recombination

There are two classes of recombination processes: *electronic recombination*, involving charge neutralization of an electron-ion pair and *ion-ion recombination* where the charge neutralization occurs as the result of a reaction between a positive and a negative ion. Two types of electronic recombination [98] can occur in a planetary ionosphere; the one involving atomic ions is called radiative recombination, while the one for molecular ions is called dissociative recombination.

Radiative Recombination

At low electron densities, and in the absence of molecular neutral species, i.e., at high altitudes in a planetary ionosphere, but more importantly in interplanetary space, radiative recombination process represents a chemical sink for atomic ions. This process can be expressed by

$$X^+ + e \rightarrow X^* + \hbar \nu \tag{4-12}$$

where * indicates an excited state of species X. Quantum-mechanical calculation of radiative recombination coefficients at $250\,°K$ give values for typical atmospheric constituents as shown in Table 20.

Table 20

Ion	H^+	He^+	C^+	N^+	O^+
$\alpha_r \times 10^{12}\ cm^3\ sec^{-1}$	4.8	4.8	4.2	3.6	3.7

Theoretically, a temperature dependence $\alpha_r \propto T^{-\frac{1}{2}}$ is predicted; however, a better empirical representation of this dependence is given by $\alpha_r \propto T^{-n}$, where $0.7 \leq n \leq 0.75$. Since α_r is typically of the order of $10^{-12}\ cm^3\ sec^{-1}$ at ionospheric temperatures $(1000\,°K)$, the lifetime against radiative recombination τ_r is extremely long, so that other processes will be responsible for the loss of atomic ions.

Dissociative Recombination

This process can be expressed by

$$XY^+ + e \rightarrow XY^* \rightarrow X + Y + \Delta E. \tag{4-13}$$

Dissociative recombination occurs as a result of a radiationless transition forming a quasimolecule at small internuclear distance, where the potential energy curve of XY^+ has a minimum, while that of XY^* is repulsive, leading to a separation of the atoms X and Y. This process leads to a dissociative recombination coefficient α_D which is about 10^5 greater than the radiative recombination coefficient α_r. Dissociative recombination is therefore one of the most important chemical loss processes [97]. Numerical values of α_D for important molecular ions in planetary ionospheres are shown in Table 21.

Table 21

Ion	α_D (cm^3 sec^{-1})*
CO_2^+	3.8×10^{-7}
O_2^+	2.2×10^{-7}
N_2^+	2.9×10^{-7}
NO^+	4.5×10^{-7}
H_2^+	3×10^{-8}
H_3^+	$\sim 10^{-8}$
H_3O^+	$\sim 10^{-6}$

The dissociative recombination coefficient shows a complicated temperature dependence: according to experiments, $\alpha_D(NO^+) \propto T_e^{-\frac{1}{3}}$; $\alpha_D(N_2^+) \propto T_e^{-\frac{1}{3}}$; $\alpha_D(O_2^+) \propto T_e^{-0.7}$ and $\alpha_D(CO_2^+) \propto T_e^{-1}$, while the variation with isothermal temperature $(T_e = T_i = T)$ is $\propto T^{-1}$ for all except $\alpha_D(N_2^+)$ which is almost constant and $\alpha_D(H_3O^+) \propto T^{-2}$. Form theoretical considerations a temperature dependence between $T_e^{-\frac{1}{2}}$ to $T_e^{-1.5}$ is expected.

The dissociative recombination coefficient may also depend on the ion temperature, due to its dependence on the vibrational structure of the molecular ion.

Ion-Ion Recombination

In the lower part of planetary ionospheres (D-region) negative ions can form as the result of electron attachment. These can recombine directly with positive ions in a process, which is also called *mutual neutralization*, according to

$$X^+ + Y^- \rightarrow X^* + Y + \Delta E \tag{4-14}$$

where the positive ion may become electronically excited. The excess energy ΔE of the reaction goes into kinetic energy of the neutralized particles or in the case of molecules into vibrational or rotational excitation. The rate coefficients for ion-ion recombination (mutual neutral-

---

* For $T_e = T_i = T_n = 300\ ^\circ$K.

Table 22

	$\alpha_{mn} \times 10^7$ cm^3 sec^{-1}	
Ions	Theoretical	Experimental
$H^+ + H^-$	1.2/4.0	4.0 ± 1.8
$N^+ + O^-$	1.8	2.9 ± 1
$O^+ + O^-$	1.1	2.8 ± 1

ization) α_{mn} are of the order of 10^{-7} cm^3 sec^{-1} [99]. Table 22 shows values of α_{mn} at $T = 300\,°$K for some constituents for which experimental data, though at higher temperature, are available. A temperature dependence of $\alpha_{mn} \propto T^{-\frac{1}{2}}$ is predicted theoretically for low energies.

IV.3. Charge Exchange Reactions

Ion-neutral (or *ion-molecule*) reactions lead to the exchange of charge between these species as the result of collisions [100]. Charge exchange reactions include *charge transfer* processes in which one or more electrons are transferred, as well as *ion-atom interchange* (charged rearrangement) processes in which a charged or neutral atomic or molecular species is transferred similar to an ordinary chemical reaction. Charge exchange reactions play an important role in planetary ionospheres where atomic ions and neutral molecular species are present, since in this case they represent the controlling chemical process, due to their rather high rate coefficient. The rate coefficient k can be defined as

$$k = \int_0^\infty v\sigma(v)\, f(v)\, dv = \langle \sigma v \rangle \qquad (4\text{--}15)$$

where v is the relative velocity of the colliding particles, σ is the reaction cross section and $f(v)$ is a normalized velocity distribution function. Theoretical estimates for charge exchange reaction rate coefficients can be made on the basis of the Langevin-Eyring-Gioumousis and Stevenson theory [96, 100] which yields a reaction cross section based on the ion-induced dipole interactions

$$\sigma = \pi e \left(\frac{2 m_i \alpha}{\mu E_i} \right)^{\frac{1}{2}} \qquad (4\text{--}16)$$

where m_i is the ion mass, $\mu = m_i m_n / (m_i + m_n)$ is the reduced mass of the ion-neutral system, α is the polarizability of the neutral species and E_i is the ion kinetic energy and e is the electronic charge.

Since the velocity of the ion is $v_i = (2E_i/m_i)^{\frac{1}{2}}$, the rate coefficient, according to (4–15), is given, using (4–16), by

$$k = 2\pi e \left(\frac{\alpha}{\mu}\right)^{\frac{1}{2}} = 2.3 \times 10^3 \left(\frac{\alpha}{\mu}\right)^{\frac{1}{2}} \text{ cm}^3 \text{ sec}^{-1}. \qquad (4\text{–}17)$$

Typical values of k based on the Gioumousis-Stevenson formula are of the order 10^{-9} cm^3 sec^{-1}, representing an upper limit. Based on the energy dependence of $\sigma (\propto E^{-\frac{1}{2}})$, the rate coefficient k should be independent of temperature; however, some experimental data suggest a temperature dependence $T^{-\frac{1}{2}}$, corresponding to a dependence of σ upon E^{-1}.

It has also been suggested, that ion-neutral reactions may involve an activation energy, which in case of an otherwise endothermic reaction, must be at least equal to the endothermicity.* In this case the temperature dependence of the rate coefficient can be represented by the Arrhenius relation [96, 101]

$$k = A \exp\left[-\frac{E_a}{k T}\right], \qquad (4\text{–}18)$$

where E_a is the activation energy (in eV), k the Boltzmann constant, and A is the so-called frequency factor. Although A is often considered to be a temperature-independent factor (following the Arrhenius theory), it may also be dependent on temperature, since it corresponds to a reduced collision frequency, which may be energy dependent, according to

$$A \equiv v'_\mu = 2\sigma (2\pi k T \mu)^{\frac{1}{2}} \qquad (4\text{–}19)$$

where μ is the reduced mass. (A is also sometimes referred to as the steric hindrance factor, borrowed from chemical reaction theory) [101.]

Charge Transfer Reactions

Reactions of this type proceed according to the scheme

$$X^+ + Y \rightarrow Y^+ + X. \qquad (4\text{–}20)$$

When the neutral species is a molecule, *dissociative charge transfer* can also take place, according to

$$X^+ + YZ \rightarrow Y^+ + Z + X. \qquad (4\text{–}21)$$

---

* The energy defect ΔE of the reaction, $X^+ + Y \rightarrow Y^+ + X + \Delta E$ representing the difference in the ionization potentials of the reaction partners, determines if the reaction is possible (exothermic). $\Delta E = \text{IP}(X) - \text{IP}(Y) > 0$ represents an *exothermic*, $\Delta E < 0$ an *endothermic* and $\Delta E = 0$ a *resonant* reaction.

A special case is the *resonant (symmetric) charge transfer*

$$X^+ + X \leftrightarrows X + X^+. \tag{4-22}$$

This reaction can be important for converting fast ions into thermal ions and at the same time producing fast neutrals. The cross section for resonant charge transfer at low energies has an energy dependence

$$\sigma^{\frac{1}{2}} = a - b \log E$$

i. e., is largest at thermal energies. In addition to the resonant charge transfer involving like ion and neutral species, there is an *accidentially resonant*, but asymmetric, *charge transfer* process, when the difference in the ionization potentials (energy defect) is very small [102]. The classic example of this type of reaction, which is of great importance in the terrestrial ionosphere, is [103]

$$O^+ + H \rightleftarrows O + H^+. \tag{4-23}$$

The equilibrium concentrations associated with this process are given by [104]

$$\frac{n(H^+)}{n(O^+)} = \frac{9\,n(H)}{8\,n(O)} \tag{4-23a}$$

Table 23

Reaction	$k\,(\mathrm{cm^3\,sec^{-1}})$
$H^+ + H \rightarrow H + H^+$	$3\,(-9)\star$
$H^+ + O \rightarrow H + O^+$	$4\,(-10)$
$H^+ + CO_2 \rightarrow H + CO_2^+$	$1\,(-10)$
$He^+ + O_2 \rightarrow He + O + O^+$	$2\,(-9)$
$He^+ + CO_2 \rightarrow O^+ + CO + He$	$1\,(-9)$
$He^+ + CO \rightarrow C^+ + O + He$	$2\,(-9)$
$He^+ + N_2 \rightarrow He + N_2^+$	$2\,(-9)$
$He^+ + N_2 \rightarrow He + N + N^+$	$8\,(-10)$
$N^+ + O_2 \rightarrow N + O_2^+$	$5\,(-10)$
$O^+ + O_2 \rightarrow O + O_2^+$	$2\,(-11)$
$N_2^+ + O_2 \rightarrow N_2 + O_2^+$	$1\,(-10)$
$N_2^+ + CO \rightarrow N_2 + CO^+$	$7\,(-11)$
$N_2^+ + NO \rightarrow N_2 + NO^+$	$5\,(-10)$
$N_2^+ + CO_2 \rightarrow N_2 + CO_2^+$	$9\,(-10)$
$CO^+ + O_2 \rightarrow CO + O_2^+$	$2\,(-10)$
$CO^+ + CO_2 \rightarrow CO_2 + CO_2^+$	$1\,(-9)$
$CO_2^+ + O_2 \rightarrow CO_2 + O_2^+$	$1\,(-10)$
$CO_2^+ + H \rightarrow H^+ + CO_2$	$1\,(-10)$

$\star\ (-X) \equiv 10^{-X}$.

where $\frac{9}{8}$ is the ratio of the products of the statistical weights obtained from detailed balancing* of reaction (4–23). A cross section of 7.6×10^{-16} cm^2 has been derived for this reaction from ionospheric data, leading to a rate coefficient $k \cong 3.9 \times 10^{-10}$ cm^3 sec^{-1} [105].

Some charge transfer reactions of importance in planetary iono- spheres and their measured or theoretically estimated rate coefficients [98, 100] are listed in Table 23.

Ion-Atom Interchange Reactions

In contrast to straightforward charge transfer, these reactions also in- volve the transfer of a charged or neutral atom according to

$$X^+ + YZ \to X Y^+ + Z. \qquad (4\text{–}24)$$

This type of reaction is sometimes called *charged rearrangement*. Direct charge transfer between atomic ions and molecular species, as well as ion-atom interchange represent important loss processes for atomic ions, since the molecular ions formed by this reaction will be subject to (fast) dissociative recombination. This represents an efficient removal path for atomic ions in planetary ionospheres [97, 107]. In altitude regions where these processes occur, the loss of atomic ions will be

Table 24

Reaction	k (cm^3 sec^{-1})
$H^+ + CO_2 \to COH^+ + O$	6 (−10)
$H_2^+ + H_2 \to H_3^+ + H$	2 (−9)
$He^+ + H_2 \to HHe^+ + H$	<(−13)
$C^+ + CO_2 \to CO^+ + CO$	2 (−9)
$C^+ + O_2 \to CO^+ + O$	1 (−9)
$N^+ + O_2 \to NO^+ + C$	5 (−10)
$N^+ + CO_2 \to NO^+ + C$	2 (−11)
$O^+ + N_2 \to NO^+ + N$	1 (−12)
$O^+ + CO_2 \to O_2^+ + CO$	1 (−9)
$O_2^+ + N \to NO^+ + O$	2 (−10)
$CO_2^+ + H \to COH^+ + O$	6 (−10)
$CO_2^+ + O \to O_2^+ + CO$	2 (−10)

---

* The statistical weights $g_s = \sum (2J+1)$ for the reactants O^+, H, H^+ and O have values 4, 2, 1 and 9, respectively. However since the reaction is not truly resonant, i. e., $\Delta E = IP(O) - IP(H) = 0.02$ eV, the ratio of the products of the statistical weights should be modified by a factor $\exp(\Delta E / \ell T)$. While this leads to a ratio $\frac{9}{8}$ for $T \to \infty$, at $T_s = 1000$ °K this ratio is $\frac{9}{7}$ and at $T_s = 600$ °K it is $\frac{7}{5}$ [106].

governed by a linear loss law. Similarly to the other charge exchange reactions, the temperature dependence of the rate constants for ion-atom interchange is often quite complex and may involve an activation energy [cf. 100].

Table 24 lists some important ion-atom interchange reactions and their rate constants [100, 107].

At altitudes where water vapor molecules are present, some ions form hydrates according to [97, 108]

$$P^+ + H_2O + Z \rightarrow P^+ \cdot (H_2O) + Z \qquad (4\text{-}25)$$

where P^+ represents O_2^+, NO^+, CO_2^+.

Repeated reactions of this type (with a three-body rate coefficient $k_3 \approx 10^{-28}$ cm^6 sec^{-1}) lead to *cluster ions* of the form

$$P^+ \cdot (H_2O)_{n-1} + H_2O + Z \rightarrow P^+ \cdot (H_2O)_n + Z. \qquad (4.26)$$

The three-body reaction chain is terminated by an exothermic binary reaction $(k \approx 10^{-10}$ cm^3 sec$^{-1})$ of the type

$$P^+ \cdot (H_2O)_n + H_2O \rightarrow H_3O^+ \cdot (H_2O)_{n-1} + P + OH \qquad (4\text{-}27)$$

forming hydrated hydronium ions (H_3O^+). Such heavy cluster ions appear to be the predominant ion species in the D region of a planetary ionosphere [108]. Similar cluster ions are formed with CO_2, leading to $P^+ \cdot (CO_2)_n$, where P^+ may also be CO_2^+.

IV.4. Negative Ion Reactions

Negative ions can be formed by attachment of electrons to neutral species. Since this process depends on the collisions between electrons and neutral atoms and molecules, formation of negative ions will occur predominantly in the densest part of a planetary ionosphere (D-region). The stability of a negative ion depends on the *electron affinity* (i. e., the binding energy of the extra electron, typically of the order 1 eV). The following attachment processes are possible [97, 109, 110]:

1) *radiative attachment*

$$e + \begin{cases} X \\ XY \end{cases} \rightarrow \begin{cases} X^- \\ XY^- \end{cases} + h\nu \qquad (4\text{-}28)$$

2) *dissociative attachment*

$$e + XY \rightarrow X + Y^- \qquad (4\text{-}29)$$

3) *three-body attachment*

$$e + X\,Y + Z \rightarrow X\,Y^- + Z \qquad (4\text{--}30)$$

where Z represents the third collision partner, usually a molecule, such as $X\,Y$.

Radiative and dissociative attachment processes are usually less important than three-body attachment, except in the upper-most part of the D-region where electron-neutral collisions become small. The rate coefficients for these two-body reactions is of the order $\sim 10^{-15}$ cm^3 sec^{-1}. The dominant process for negative ion formation in the lowermost part of a planetary ionosphere $(p \geq 10^{-2}$ Torr) is thought to be three-body attachment which has a rate coefficient $a_3 \cong 10^{-30}$ cm^6 sec^{-1}.

Although the various attachment processes are either two- or three-body reactions, they are often represented by a linear loss law appropriate to a one-body reaction, according to*

$$\frac{dN_e}{dt} = -\beta_a N_e \qquad (4\text{--}31)$$

where β_a is the rate of attachment per free electron (sec^{-1}), given by the rate coefficient multiplied by the number density of the neutral species for the two-body processes and multiplied by the square of the neutral density for the three-body process. The reciprocal of β_a is the lifetime of an electron against attachment τ_{att} (sec).

Destruction of negative ions occurs by *detachment* processes and ion-ion recombination (mutual neutralization; see (4–14)).

The following detachment processes can occur:
1) *photo-detachment*

$$X\,Y^- + h\nu \rightarrow X\,Y + e \qquad (4\text{--}32)$$

2) *associative detachment*

$$X\,Y^- + Z \rightarrow X\,YZ + e \qquad (4\text{--}33)$$

3) *Penning detachment*

$$X\,Y^- + X\,Y^* \rightarrow X\,Y + X\,Y + e \qquad (4\text{--}34)$$

which involves an excited (metastable) species $X\,Y^*$.

The last two types of attachment processes are also categorized as *collisional detachment*.

---

* This is the reason why in the older ionospheric literature, the linear loss law in the F region which results from a completely different reaction (charge exchange), is sometimes referred to as "quasi-attachment law".

Again, photo-detachment appears to be of lesser importance than the collisional detachment processes. The latter have rate coefficients of the order $10^{-10}\,\mathrm{cm^3\,sec^{-1}}$; photo-detachment rates are of order $0.3\,\mathrm{sec^{-1}}$ [109, 110].

Analogously to charge exchange processes for positive ions, *negative ion-molecule reactions* lead to the transformation of one negative ion species into another [97]. Basically these are simple charge transfers of the type

$$X^- + Y \rightarrow X + Y^- \tag{4-35}$$

where X and Y are usually molecular species, and negative ion-molecule interchange,

$$X^- + YZ \rightarrow XY^- + Z \tag{4-36}$$

which can also occur as a three-body process. Examples of negative ion-molecule reactions [108] and their rate coefficients are listed in Table 25.

Table 25

Reaction	Rate coefficient $(\mathrm{cm^3\,sec^{-1}})$
$O_2^- + O_3 \rightarrow O_3^- + O_2$	$3\,(-10)$
$O_2^- + NO_2 \rightarrow NO_2^- + O_2$	$8\,(-10)$
$O_3^- + NO \rightarrow NO_3^- + O$	$1\,(-11)$
$O_3^- + CO_2 \rightarrow CO_3^- + O_2$	$4\,(-10)$
$CO_3^- + NO \rightarrow NO_2^- + CO_2$	$9\,(-12)$
$CO_3^- + O \rightarrow O_2^- + CO_2$	$8\,(-11)$
$NO_2^- + O_3 \rightarrow NO_3^- + O_2$	$2\,(-11)$
$O_2^- + O_2 + Z \rightarrow O_4^- + Z$	$>(-30)\,[\mathrm{cm^6\,sec^{-1}}]$
$O_4^- + O \rightarrow O_3^- + O_2$	$4\,(-10)$
$O_4^- + CO_2 \rightarrow CO_4^- + O_2$	$4\,(-10)$
$CO_4^- + O \rightarrow CO_3^- \pm O_2$	$2\,(-10)$
$CO_4^- + NO \rightarrow NO_3^- + CO_2$	$5\,(-11)$
$CO_3^- + NO_2 \rightarrow NO_3^- + CO_2$	$8\,(-11)$

The condition of charge neutrality requires that

$$N_+ = N_e + N_-. \tag{4-37}$$

Whenever negative ions are present, their effect on the ionization balance is expressed in terms of a parameter $\lambda^- = N_-/N_e$, and thus $N_+ = (1 + \lambda^-) N_e$. The parameter λ^- depends on the rate coefficients for the various processes involved in the formation and destruction of negative ions, according to

$$\lambda^- \simeq \frac{v_{att}}{v_{coll.\,det.} + v_{ph.\,det.} + v_{mn}} \qquad (4\text{--}38)$$

where the v_s are the frequencies per ion or electron for the appropriate processes, i. e., the rate coefficients times the number density, according to a linear (one-body) rate equation.

With positive and negative ions present, the effective recombination coefficient is

$$\alpha_{eff} = \alpha_e + \lambda^- \alpha_i \qquad (4\text{--}39)$$

where α_e is the electronic and α_i the ionic recombination coefficient. The following Table 26 lists negative ions identified by mass spectrometers in the terrestrial ionosphere.

Table 26

Negative Ion	Mass Number
O_2^-	32
Cl^-	35, 37
CO_3^-	60
HCO_3^-	61
NO_3^-	62
$O_2^- \cdot (H_2O)_2$	68
CO_4^-	74
$CO_3^- \cdot H_2O$	78
$NO_2^- \cdot (HNO_2)$	93 ± 1
$CO_4^- \cdot (H_2O)$	104
$NO_2^- \cdot (HNO_2) \cdot H_2O$	111 ± 1
$CO_4^- \cdot (H_2O)_2$	122
$NO_3^- \cdot (HNO_3)$	125 ± 1

IV.5. Chemical Reactions and Airglow

Recombination processes provide a source of airglow emission, which is particularly observable during the night *(nightglow)* [56, 111]. Radiative recombination of atomic ions leads to emission according to

$$X^+ + e \rightarrow X^* + h\nu \qquad (4\text{--}40)$$

which can be followed by

$$X^* \rightarrow X \text{ (ground state} + \sum h\nu)$$

where $\hbar v$ is the direct recombination photon and $\sum \hbar v$ represent the photons emitted in cascade to the ground state.

The emission rate from radiative recombination can be expressed by [111]

$$\mathscr{I}_\lambda = k_\lambda \int\limits_{z_1}^{z_2} \alpha_r N_i N_e \, dz = k_\lambda \int\limits_{z_1}^{z_2} \alpha_r N^2 \, dz \qquad (4\text{-}41)$$

where k_λ is the probability that line λ will be emitted from a given recombination, α_r is the radiative recombination coefficient and N_i and N_e are the ion and electron density, respectively; if the ion represents the predominant ion, $N_i = N_e \equiv N$, in the altitude range between z_1 and z_2. It has been suggested [112, 113], that nightglow emissions in the terrestrial ionosphere may be the result of the raction

$$O^+(^4S) + e \rightarrow O^* + \hbar v$$
$$O^* \rightarrow O(^3P) + \sum \hbar v. \qquad (4\text{-}42)$$

The integrated (column) emission rate in the vertical direction from an energy level j, which is populated by $[X_j^*]$ atoms per cm^3, can be expressed by

$$\mathscr{I}_\lambda = A_\lambda \int\limits_0^\infty [X_j^*] \, dz \qquad (4\text{-}43)$$

where A_λ is the Einstein transition coefficient.

The concentration $[X_j^*]$ is governed by an equation of continuity, where the production is given by radiative transitions, while the loss is primarily due to collisional deactivation ("quenching") [111].

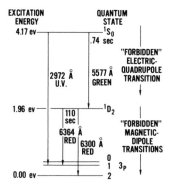

Fig. 36. Transitions in atomic oxygen leading to the 5577Å and 6300Å airglow emissions

Dissociative recombination usually provides for a substantial energy defect ΔE

$$X Y^+ + e \rightarrow X + Y + \Delta E.$$

This type of reaction allows the dissociation products to be energetically in excited levels; transition to the ground level (or intermediate levels) then leads to the emission of radiation which is most easily observed as nightglow. The emission rate can be expressed analogously to (4–41).

Typical examples of nightglow emissions in the terrestrial ionosphere, associated with dissociative recombination are the reactions

$$O_2^+ + e \rightarrow O + O + 6.96 \, eV$$

$$NO^+ + e \rightarrow N + O + 2.76 \, eV$$

which lead to the emission of the 5577 Å green line of oxygen and the 6300 Å red line of oxygen. The typical O transitions are shown in Fig. 36.

Dissociative recombination of CO_2^+, according to reaction

$$CO_2^+ (\tilde{X}^2 \, \Pi_g) + e \rightarrow CO(a^3 \, \Pi) + O$$

may also contribute to the observed emission of the $CO(a-X)$ Cameron band (1900—2500 Å) in the atmosphere of Mars [57].

Chapter V

Plasma Transport Processes

Plasma transport can act as an apparent local source or sink in a planetary ionosphere. Transport processes occur as the result of pressure gradients and gravity (plasma diffusion), the action of an electric field (plasma drift) or as the result of frictional forces (plasma wind).

V.1. Plasma (Ambipolar) Diffusion

In a weakly ionized plasma, such as a planetary ionosphere, electrons and ions diffuse through the ambient atmosphere, similar to minor constituents in the neutral atmosphere. However, since electrons, because of their higher mobility, diffuse faster than the ions, a polarization field is set up which prevents charge separation. As the result of this polarization field E, the electrons and ions diffuse with the same velocity. This process is called *ambipolar* diffusion [4]. The plasma diffusion velocity $v_D = v_e = v_i$ can be derived (for the steady state) from the equations of motion for the electrons and ions considering the effects of their partial pressure gradients ∇p and the effect of gravity, g, according to

$$\nabla p_e = -\rho_e g - q_e E - \rho_e v_{en}(v_e - v_n)$$
$$\nabla p_i = -\rho_i g + q_i E - \rho_i v_{in}(v_i - v_n)$$

(5–1)

with $p_j = N_j k T_j$, $\rho_j = N_j m_j$ and $q_j = |e| N_j$.

Assuming the neutral atmosphere at rest, $v_n = 0$ and taking into account $m_i \gg m_e$ and $m_i v_{in} \gg m_e v_{en}$, i.e., ion-neutral collisions to be more important than electron-neutral collisions, we obtain for the plasma (ambipolar) diffusion velocity

$$v_D = -D_a \left[\frac{\nabla p_N}{p_N} + \frac{1}{\mathcal{H}} \right]$$

(5–2)

where the plasma pressure $p_N = N k (T_e + T_i)$ and the plasma scale height

$$\mathcal{H} = \frac{k(T_e + T_i)}{m_i g}$$

(5–3)

and the ambipolar diffusion coefficient

$$D_a = \frac{\mathscr{k}(T_e + T_i)}{m_i v_{in}}. \tag{5-4}$$

(For $T_e = T_i$, $D_a = 2D_i$; i. e., twice the ion diffusion coefficient.) The ion-neutral collision frequency can be expressed for $T < 2000\,°K$ by [cf. (3–24)]

$$v_{in} \cong 2\pi \left(\frac{\alpha e^2}{\mu_{in}}\right)^{\frac{1}{2}} n$$

where α is the polarizability, e is the ionic charge, $\mu_{in} = m_i m_n/(m_i + m_n)$ is the reduced mass and n is the number density of the neutrals.

For diffusion of ions in their parent gas (X^+ in X), the ion-neutral collision frequency is modified due to resonance charge transfer [90] introducing a dependence on the neutral and ion temperature

$$v_{in} \propto (T_i + T_n)^m$$

where $0.3 \lesssim m \lesssim 0.4$ for most atmospheric constituents, so that

$$D_a(X^+, X) \propto (T_e + T_i)(T_i + T_n)^{-m}. \tag{5-5}$$

Typical ambipolar diffusion coefficients at $T_e = T_i = T_n = 1000\,°K$ for ions in their parent gas are listed below:

$$D_a(O^+, O) = \frac{10^{19}}{n(O)} \text{ cm}^2 \text{ sec}^{-1}$$

$$D_a(H^+, H) = \frac{2.5 \times 10^{19}}{n(H)} \text{ cm}^2 \text{ sec}^{-1}.$$

Since the ambipolar diffusion coefficient D_a is inversely proportional to the density of the (neutral) constituent through which the electron-ion gas diffuses, it can also be expressed by

$$D_a = D_0 \exp\left(\frac{z}{H_D}\right) \tag{5-6}$$

where $H_D = \mathscr{k}T_j/m_j g$ is the scale height of the constituent through which ambipolar diffusion takes place.

For the case of an isothermal atmosphere ($\mathscr{H} = \text{const}$), the diffusion velocity has the form

$$v_D = -D_a \left[\frac{1}{N}\frac{\partial N}{\partial z} + \frac{1}{\mathscr{H}}\right]. \tag{5-7}$$

When a planetary magnetic field is present, plasma diffusion is constrained along $\boldsymbol{B}$. The diffusion velocity in the vertical (z) direction is then given by

$$v_z = v_D \sin^2 I \qquad (5\text{–}8)$$

where I is the inclination (dip) of the magnetic field **B** (with respect to the surface)* [114, 115].

The diffusion velocity enters into the ionospheric continuity equation in form of the divergence of a flux, $\mathbf{V}\cdot(N\,v_D)$. For the simple case of an isothermal atmosphere $H=$const. $T_e=T_i=T_n$ and $m_i=m_n$, representing diffusion of an ion through its parent gas**, $\mathscr{H}=2H$, we obtain

$$\mathbf{V}\cdot(N\,v_D) \simeq \frac{\partial}{\partial z}(N v_D) = D_a\left\{\left(\frac{\partial N}{\partial z}\right)^2 + \frac{3}{2}\frac{\partial N}{\partial z} + \frac{N}{2H^2}\right\}. \qquad (5\text{–}9)$$

The condition for diffusive equilibrium [116] can be expressed by

$$\mathbf{V}\cdot(N\,v_D) \simeq \frac{\partial(N v_D)}{\partial z} = 0. \qquad (5\text{–}10)$$

Equation (5–10) can be satisfied uniquely by two solutions:

a) $N v_D = $ const,

b) $v_D = 0$. $\qquad (5\text{–}11)$

Of these solutions a) is not a true equilibrium condition since it corresponds to a constant flux $F=Nv$ leading to an outflow of ionization to a sink at infinity. This solution, however, could be considered as a *dynamic* equilibrium. Solution b) represents a true *(static)* equilibrium, i. e., the solution where the plasma has reached a stable distribution as the result of ambipolar diffusion. The time constant for reaching diffusive equilibrium is given by

$$\tau_D \simeq \frac{H_D^2}{D_a} \qquad (5\text{–}12)$$

which can be obtained from the solution of

$$\frac{\partial N}{\partial t} = -\mathbf{V}\cdot(N\,v_D)$$

---

* The reason for the $\sin^2 I$ factor is as follows: Ambipolar diffusion under gravity in the presence of an inclined magnetic field leads to a diffusion velocity along B, $v_\parallel=v_D\sin I$; the z-component is $v_z=v_\parallel\sin I$; thus $v_z=v_D\sin^2 I$.
** The term effective plasma temperature is sometimes employed according to $T_p'=(T_e+T_i)/2\cong(T_e+T_n)/2$.

for condition of maximum flux (see below) and $\mathcal{H} > H_D$, noting that

$$\tau \equiv \left(\frac{1}{N}\frac{\partial N}{\partial t}\right)^{-1}.$$

The general solution of (5–10) is given by

$$N = A\,e^{-\frac{z}{\mathcal{H}}} + B\,e^{-\frac{z}{H_D}} \tag{5–13}$$

where the appropriate boundary conditions

$$N = 0 \quad \text{at } z \to \infty \quad \text{and} \quad N = N_0 \quad \text{at } z \to 0$$

lead to

$$F = N v_D = \frac{D_0 B(\mathcal{H} - H_D)}{H_D \mathcal{H}}$$

$$B = N_0 - A \tag{5–14}$$

$$A = N_0 - \frac{F H_D \mathcal{H}}{(\mathcal{H} - H_D)D_0}.$$

$B=0(A=N_0)\to F=0$, represents the diffusive equilibrium solution b), whereas $A=0$ represents the *maximum upward flux* for steady state conditions, i. e.,

$$F^* = \frac{D_0 N_0}{H_D \mathcal{H}}(\mathcal{H} - H_D). \tag{5–15}$$

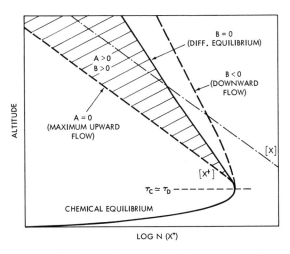

Fig. 37. Chemical equilibrium, diffusive equilibrium and the effects of flow (i. e., ambipolar diffusion through the neutral gas X) for a major ion X^+

$B>0$ represents an *upward* flux and $B<0$ a *downward* flux of ionization. The solution for maximum upward flux $(A=0)$ corresponds to a plasma density distribution having a scale height H_D, representing a constituent through which the plasma diffuses, similar to the diffusion of minor neutral constituents (cf. I.2). This is illustrated in Fig. 37.

The plasma density distribution in the presence of flow can also be written as

$$N = N_0 e^{-\frac{z}{\mathscr{H}}}\left[1 - \frac{F H_D \mathscr{H}}{N_0 D_0 (\mathscr{H} - H_D)}\left(1 - e^{-\left(\frac{\mathscr{H}-H_D}{\mathscr{H}H_D}\right)z}\right)\right] \qquad (5\text{--}16)$$

where the first term represents the diffusive equilibrium distribution and the second term the modification due to a flux F. For the maximum flux $F^* = (D_0 N_0 / H_D \mathscr{H})(\mathscr{H} - H_D)$, the factor of the second term becomes unity, leading to a distribution $N = N_0 e^{-z/H_D}$. This is identical to the situation discussed for the diffusion of a minor neutral constituent (I.2).

The diffusive equilibrium distribution for electron and ions can also be derived directly from equations (5–1) neglecting all terms containing velocities. If one considers more than one ionic species, one obtains

$$\frac{1}{p_N}\frac{\partial p_N}{\partial z'} = -\frac{m_+ g}{k(T_e + T_i)} \equiv -\frac{1}{\mathscr{H}} \qquad (5\text{--}17)$$

where $m_+ = \sum \rho_i / N$ is the mean ionic mass and $p_N = k(T_e + T_i)$ is the plasma pressure, z' is the reduced altitude $\left(z' = \int_0^z (g/g_0)dz\right)$ accounting for the height variation of the acceleration of gravity. However, plasma pressure is not a directly observable quantity, but plasma density and temperatures are generally measured.

The plasma density distribution can be defined in terms of a density scale height H'_N

$$\frac{1}{N}\frac{\partial N}{\partial z'} \equiv \frac{\partial(\ln N)}{\partial z'} = -\frac{1}{H'_N}$$

where $(5\text{--}18)$

$$H'_N = \mathscr{H}\left[1 + \mathscr{H}\frac{\partial(T_e + T_i)/\partial z'}{T_e + T_i}\right]^{-1}$$

and $H_N \equiv \mathscr{H}$ only when T_e, $T_i = \text{const.}$

In contrast to the behavior of neutral constituents in diffusive equilibrium, the diffusive equilibrium distribution of a particular ion X^+ in an ion mixture having a mean ionic mass m_+ is *not* independent of the other ions [117, 118]. This is due to the existence, in addition to gravity, of the polarization field E in a plasma which acts on all ions and which depends on the mean ionic mass and the charged particle temperatures. The distribution of an ion species X^+ in diffusive equilibrium $(v_e = v_i = 0)$ can be derived from eqs. (5–1) together with the condition for charge neutrality★ $\sum N_i = N_e$ and the definition of mean ionic mass $m_+ = \sum \rho_i/N$, and is given by [119]

$$N_i(X^+) = N_{i0} \exp\left\{ -\int_0^z \left[\left(m(X^+) - \frac{m_+ T_e}{T_e + T_i}\right) \frac{g}{\mathscr{k} T_i} + \frac{\partial(T_e + T_i)/\partial z}{T_e + T_i} \right] dz \right\}.$$

(5–19)

The electrostatic polarization field

$$eE = \frac{m_+ g T_e}{T_e + T_i}$$

(5–20)

counteracts the effect of gravity leading to an increase in the concentration of a light minor ion X^+ (neglecting temperature gradients) while in diffusive equilibrium, as long as

$$\hat{m}(X^+) < \frac{\hat{m}_+ T_e}{(T_e + T_i)}.$$

The effect of this polarization field is to give minor light ions a substantial effective upward acceleration, e. g., for H^+ in a CO_2^+ ionosphere this acceleration is $(\hat{m}(X^+) - \hat{m}_+/2)g \cong 21\,g$, assuming $T_e = T_i$. The distribution of minor ions H^+ and He^+ in an O^+ ionosphere is shown in Fig. 38.

While the altitude distribution of individual ions is also dependent on the electron temperature, as the result of the effect of the polarization field, the *relative* abundance (ratio) of ionic species in diffusive equilibrium depends on the ion temperature only, since the terms containing T_e cancel. Thus,

$$\frac{N_i(X^+)}{N_i(Y^+)} = \frac{N_0(X^+)}{N_0(Y^+)} \exp\left(\frac{z}{H_{XY}}\right)$$

(5–21)

where

$$H_{XY} = \frac{\mathscr{k} T_i}{(m(X^+) - m(Y^+))g}.$$

★ It can be shown that $\nabla \cdot E \approx 0$, i. e., the charge excess due to the polarization field E is $(N_i - N_e)/N_e \simeq Gm_i^2/2e^2 \simeq 4 \times 10^{-37}$ (for $T_e = T_i$ and $m_i = m_H$), where G is the universal gravitational constant and e is the electronic charge [137]. This result follows from $\nabla \cdot E = -\phi_E \cong (m_i/2e)\Delta\phi_g$, where ϕ_E is the electric and ϕ_g the gravitational potential (see VI.5), satisfying the Poisson equation $\Delta\phi_E = -4\pi e \sum_j Z_j N_j$ and $\Delta\phi_g = 4\pi G \sum_j N_j m_j$.

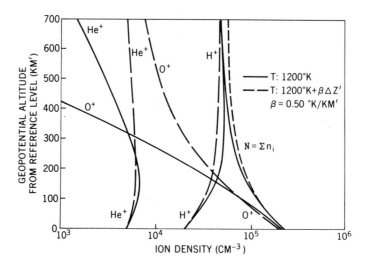

Fig. 38. Diffusive equilibrium distributions for minor ions H^+ and He^+ in an O^+ ionosphere for isothermal $(T_e = T_i = \text{const})$ conditions and for a temperature gradient $dT_e/dz = dT_i/dz = \beta = \text{const}$

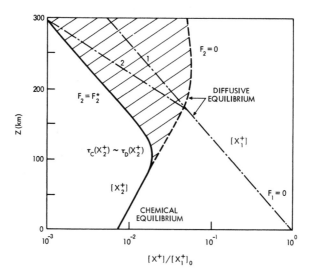

Fig. 39. Altitude distribution for a minor ion $(X_2^+ = H^+)$ in chemical and diffusive equilibrium and under flow conditions in an ionosphere with a major ion $(X_1^+ = O^+)$; distribution of X_1^+ labeled 1 corresponds to $m_+ \approx m(X_1^+)$, while 2 corresponds to $m_+ \approx m(X_2^+)$

Similarly to the case illustrated for the electron (major ion) density distribution, fluxes also modify the diffusive equilibrium distribution of a minor ion. E.g., a large upward flux will cause the distribution of the minor ion to follow approximately the scale height of the major ion through which it diffuses* (see Fig. 39).

When gradients in the charged particle temperature are present the process of *thermal diffusion* becomes also possible [120, 121]. This effect can be expressed, for a minor ion N_j, through the diffusion equation

$$\frac{1}{N_j}\frac{\partial N_j}{\partial z} = -\frac{m_j g}{\mathscr{k} T_i}-\frac{T_e}{T_i}\frac{1}{N}\frac{\partial N}{\partial z}-\frac{(1-\alpha_j)}{T_i}\frac{\partial T_i}{\partial z}-\frac{1}{T_i}\frac{\partial T_e}{\partial z} \qquad (5\text{-}22)$$

where α_j is the thermal diffusion coefficient which depends on the ratio m_j/m_+ according to Table 27.

Table 27

$\dfrac{m_j}{m_+}$	α_j
16	2.4
4	1.7
0.25	−1.1
0.0625	−1.2

Since the effective polarization field changes due to thermal diffusion according to

$$e\mathbf{E} = -\frac{\mathscr{k} T_e}{N}\mathbf{\nabla} N-(1+\alpha)\mathscr{k}\mathbf{\nabla} T_e \qquad (5\text{-}23)$$

the altitude distribution of a minor ion is affected by thermal diffusion.

Changes in effective scale height of ions in the terrestrial ionosphere at 700 km for a temperature gradient $\partial T_i/\partial z = 4\,°\text{K km}^{-1}$ and thermal diffusion are illustrated in Table 28 [120].

Table 28

	$\dfrac{10^8}{N}\dfrac{\partial N}{\partial z}\ \text{cm}^{-1}$ (at 700 km)		
	O^+	He^+	H^+
with thermal diffusion	−6.0	−0.99	+8.74
without thermal diffusion	−6.0	+2.1	+4.2

---

* The mutual diffusion coefficient for ions is given by [4]

$$D_{12} \approx \frac{10^{-21}\cdot T_i^{\frac{3}{2}}}{\sqrt{\mu_{12}}(N_1+N_2)}.$$

While thermal diffusion has important consequences for the distribution of minor ions, its effect is negligible for the major ion (electron) density distribution.

The above discussion has assumed that no planetary magnetic field is present. If, however, such a field is present, then the distributions apply *along* a dipole field line [123, 124]. Similarly, the equations are modified for multiply charged ions [118, 122] since in the foregoing discussion we have tacitly assumed that all ions are *singly charged*.

V.2. Plasma Drift and Winds

In addition to diffusion resulting from the action of pressure gradients and gravity, electrodynamic forces and neutral winds affect the ionospheric plasma [125]. The time constants controlling these motions are the collision times

$$\tau_{jn} = \frac{m_j}{\mu_{jn} \nu_{jn}} \tag{5-24}$$

where m_j is the mass of the ions, μ_{in} is the reduced mass of the collision partners, and the gyration time in the magnetic field

$$\tau_{Bj} = \Omega_{Bj}^{-1} \tag{5-25}$$

where $\Omega_{Bj} = eB/m_j c$ is the ion gyrofrequency. The equation of motion, ignoring the force terms due to gravity and pressure gradients can be written

$$\frac{\partial v}{\partial t} \cong \frac{v}{\tau} = \frac{eE}{m_j} + \frac{1}{\tau_{Bj}} \frac{(v \times B)}{c} + \frac{1}{\tau_{jn}} (v_n - v_i). \tag{5-26}$$

In general, the controlling time constants are τ_{Bj} and τ_{jn}. For $v_n \leq v_i$ and $\tau_{in} \gg \tau_{Bj}$ equ. (5–26) reduces to

$$E + \frac{1}{c} (v_j \times B) = 0. \tag{5-27}$$

According to (5–27), $v_e = v_i \equiv v_E$ represents the *electrodynamic drift velocity* which is given by

$$v_E = \frac{c(E \times B)}{B^2}. \tag{5-28}$$

The electric field E represents a polarization field which forces v_E to lie in an equipotential surface S defined by

$$\nabla S = -E = \frac{1}{c} (v_E \times B).$$

The electrons and ions are "frozen in" with the magnetic flux tube.

In the regime where $\tau_{in} \leq \tau_{Bi}$, the ions are strongly coupled with neutrals, and the ion velocity is much less than that of the electrons, leading to an electric current, according to Ohm's law

$$j = (\sigma) E . \tag{5-29}$$

Since the ionosphere in the presence of a magnetic field is highly anisotropic, the conductivity σ is expressed by a tensor.

The current flowing in an anisotropic medium, can be expressed in terms of components parallel and perpendicular to the magnetic field

$$j_{\parallel} = \sigma_0 E_{\parallel}$$
$$j_{\perp} = \sigma_1 E_{\perp} + \sigma_2 \frac{c(B \times E_{\perp})}{B^2} . \tag{5-30}$$

The conductivities are defined in this case as follows: σ_0 is the *longitudinal* conductivity which corresponds to that parallel to B or that in absence of a magnetic field, σ_1 is the *Pedersen* conductivity which is perpendicular to B, and σ_2 is the *Hall* conductivity which is perpendicular to both the magnetic and the electric field. These conductivities are defined in terms of the appropriate collision and gyrofrequencies according to

$$\left. \begin{aligned} \sigma_0 &= e^2 \sum_{i,e} \frac{N_j}{m_j v_{jn}} \simeq \frac{N_e e^2}{m_e v_{en}} \\ \sigma_1 &= e^2 \sum_{i,e} \frac{N_j}{m_j v_{jn}} \left(\frac{v_{jn}^2}{v_{jn}^2 + \omega_{Bj}^2} \right) \\ \sigma_2 &= e^2 \sum_{i,e} \frac{N_j \omega_{Bj}}{m_j(v_{jn}^2 + \omega_{Bj}^2)} . \end{aligned} \right\} \tag{5-31}$$

In the ionosphere, where the vertical scale is small compared to the horizontal scale, vertical currents are essentially zero. The horizontal current can be thought of flowing in a thin layer with the x coordinate directed southward and the y coordinate eastward according to

$$\begin{pmatrix} j_x \\ j_y \end{pmatrix} = \begin{pmatrix} \sigma_{xx} & \sigma_{xy} \\ -\sigma_{xy} & \sigma_{yy} \end{pmatrix} \begin{pmatrix} E_x \\ E_y \end{pmatrix} . \tag{5-32}$$

For magnetic dip angles $I > 3°$

$$\sigma_{xx} \simeq \frac{\sigma_1}{\sin^2 I}$$
$$\sigma_{yy} \simeq \sigma_1 \tag{5-33}$$
$$\sigma_{xy} \simeq \frac{\sigma_2}{\sin^2 I}$$

and near the magnetic equator, i.e., for $I < 3°$

$$\sigma_{xx} = \sigma_0$$

$$\sigma_{yy} = \frac{\sigma_1 + \sigma_2^2}{\sigma_1} \equiv \sigma_3 \qquad (5\text{--}34)$$

$$\sigma_{xy} = 0 .$$

At the magnetic equator, the *Cowling* conductivity σ_3 is large, leading to a large current, the so-called *equatorial electrojet*. The region where this current flows (e.g., the terrestrial E region) is also referred to as *dynamo* region, since the motion of the neutral gas by tidal and thermal forces which are coupled with the plasma lead to an induced current in the presence of a magnetic field, similar to a "dynamo". These dynamo currents are responsible for geomagnetic variations. Above the dynamo region, i.e., where $\tau_{Bi} \leq \tau_{in}$, the electrostatic polarization field generated by the dynamo currents which is communicated along magnetic field tubes, can cause a plasma drift (cf. 5–28). (This region is sometimes referred to as "*motor* region", although dynamo effects can also appear there. In the terrestrial ionosphere this corresponds to the F region.)

As the result of pressure gradients in the thermosphere, *horizontal neutral winds* are generated which can drive the ionization, provided $\tau_{Bi} \leq \tau_{in}$ [127]. In the presence of a magnetic field $\boldsymbol{B}$, the ionospheric plasma will move along magnetic field lines as the result of a neutral wind of velocity v_n, with a plasma (wind) velocity

$$v_{w\parallel} = \frac{(\boldsymbol{v_n} \cdot \boldsymbol{B})\,\boldsymbol{B}}{B^2} = v_n \cos I . \qquad (5\text{--}35)$$

The component of the plasma motion in the z direction will therefore be

$$v_{wz} = v_n \cos I \sin I \qquad (5\text{--}36)$$

which can be either upward or downward, depending on the direction of the neutral wind.

As the result of the interaction between the neutral wind and the plasma, the horizontal (x) neutral wind and plasma velocities differ by

$$v_n - v_{wx} = v_n \sin^2 I . \qquad (5\text{--}37)$$

This effect is called *air drag*, causing a dependence of the neutral gas motions on the magnetic dip angle I.

The converse situation, where neutral air is accelerated by collisions between ions and neutral particles as the result of *plasma drift* due to an electric field, $(\boldsymbol{v_i} \equiv \boldsymbol{v_E})$ can be described by

$$\frac{\partial \boldsymbol{v_n}}{\partial t} = \frac{1}{\tau_{ni}} (\boldsymbol{v_i} - \boldsymbol{v_n}) + \frac{\eta}{\rho} \frac{\partial^2 \boldsymbol{v_n}}{\partial z^2} . \qquad (5\text{--}38)$$

The acceleration of air by the plasma motion (first term) which is called *ion drag* is important as long as $\tau_{ni} = (n/N_i)\tau_{in}$ is small enough so that the motion is not hindered by viscosity (η) effects, represented by the second term.

In the steady state, ion drag causes the neutral gas to move (horizontally) with a speed $v_n = v_i \cos I$, while the ions are dragged along field-lines with a speed $v_i \cot I$ due to prevailing neutral winds.

The geometry of drift and wind effects is shown in Fig. 40.

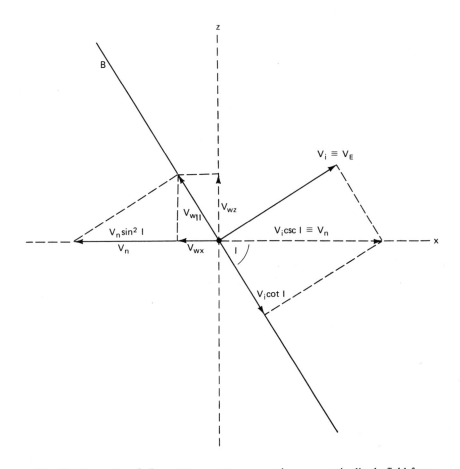

Fig. 40. Geometry of plasma transport processes in a magnetic dipole field for a dip angle I. The left side of the figure illustrates the effects of a neutral wind, while the right hand side shows the effects of a plasma drift due to an $\mathbf{E} \times \mathbf{B}$ force (with $\mathbf{E}$ perpendicular to the plane of the paper) on the ionospheric plasma and the neutral gas

V.3. Plasma Escape

In analogy with the neutral atmosphere it is possible to define an exosphere for the ionospheric plasma, satisfying the condition that the plasma scale height become comparable to the electron-ion mean free path $(\lambda_e \simeq 10^4 \, T_e^2/N_e)$. At altitudes above which this condition is fulfilled, defining an ion-exobase or baropause*, ions (and electrons) can escape from the gravitational field, provided there is no magnetic field to constrain the plasma. In the presence of a planetary magnetic field, plasma escape is possible only where the field lines are not closed loops (Fig. 41). For a dipolar magnetic field, a simple ion-exosphere model

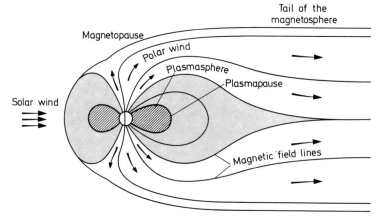

Fig. 41. Ionosphere—magnetosphere configuration for Earth showing the corotating plasmasphere and the polar wind region associated with the open geomagnetic tail. (After Banks [82])

results in a plasma density distribution along (closed) fieldlines, which decreases with altitude much more rapidly than according to a diffusive equilibrium distribution [128, 129]. By applying the law of conservation of energy together with the first magnetic invariant for the motion of charged particles in a dipole field, an idealized model for the distribution of plasma under ion-exospheric ("collisionless") conditions is given by [128]

$$N(R, L) = N_M \exp\left[\frac{-R'}{\mathscr{H}}\right]\left\{1 - \left(1 - \frac{B}{B_M}\right)^{\frac{1}{2}} \exp\left[-\frac{R'B}{\mathscr{H}(B_M - B)}\right]\right\} \quad (5\text{–}39)$$

---

* The ion-exobase occurs at a density $N_e \lesssim 10^{-4} \, (m_i/m_H) g \, T_e$, corresponding to $\mathscr{H} \simeq \lambda_e$.

where $\mathscr{H}$ is the plasma scale height, R' is a reduced altitude parameter (taking into account the variation of gravity, cf. (1–7)) along a magnetic fieldline $(R = R_0 \cos^2 \varphi)$ whose apex (at the equator) has a planetocentric distance $L = R_a/R_0 = (\cos^2 \varphi)$ with R_0 the planetary radius, and which originates at a geomagnetic latitude φ; the subscript M refers to the baropause level where a Maxwellian distribution applies. The first term of equ. (5–39) represents the diffusive equilibrium distribution, and the second term a weighting function due to the "collisionless" motion of the charged particles in the magnetic field of intensity $B(B_a \propto L^{-3})$. Comparisons between a diffusive equilibrium, an ion-exosphere and two power-law distributions* for the case of the terrestrial ionosphere are shown in Fig. 42. This type of distribution is applicable to closed

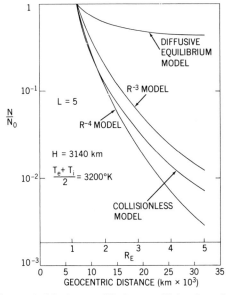

Fig. 42. Comparison of diffusive equilibrium, collision-less (ion-exosphere) and power-law distributions along a magnetic field line for $L = 5$, i.e., whose apex is at a geocentric distance of $5\,R_E$. (After Bauer [116])

dipole lines. However, as the result of the solar wind interaction with a planetary magnetosphere or ionosphere, magnetic field lines can be dragged in a tail-like formation in the antisolar direction (e.g., the geomagnetic tail) with the fieldlines parallel and essentially *open* [130].

---

* An L^{-4} power-law represents an essentially constant number of particles in a magnetic flux tube, whose volume varies as L^4, while its cross section varies as L^3.

Under these circumstances plasma can escape from the ionosphere. This escape is enhanced relative to the neutral particles as the result of the electrostatic polarization field which accelerates minor light ions and even for the major ions reduces the effect of gravity (see 5–20).

Two types of plasma escape have been suggested [131]. One is *evaporative* in nature, i.e., similar to the escape of neutrals based on the thermal energy of the plasma, leading to outflow of plasma with subsonic and sonic velocities near the exobase [132, 133], the other is based on the *hydrodynamic properties* of the plasma leading to the possibility of supersonic velocities for the outflowing plasma [134, 135]. The latter has been termed the *"polar wind"* in analogy to the solar wind, whereas the former amounts to a polar *"breeze"* only. The controversy as to the validity and relative merits of the two approaches to plasma escape has recently subsided. In fact, both approaches should lead to identical results although originally this was not the case.

The hydrodynamic approach [134, 135] is based on the bulk properties of the medium and high (supersonic) flow velocities are obtained even near the ion-exobase. As discussed before, the electrostatic polarization field in a hydrostatic (diffusive equilibrium) ionosphere $E = -d(\ln p_e)/dz = -m_+ \cdot g/2$ (for $T_e = T_i$), accelerates light ions $m_l < m_+$ upwards, leading to a massflow. Coulomb collisions between the light and major ions result in frictional drag which impedes their upward flow.

Assuming ambipolar motion $\left(N_e v_e = \sum_j N_j v_j\right)$ *along* magnetic field lines $(s)\star$ and speeds low relative to the electron thermal speed, i.e., $v_j^2 \ll \mathcal{k} T_e/m_e$, and neglecting electron collisions, the momentum equations for electrons and ions (j-th species) can be written [135]

$$\frac{1}{N_e}\frac{\partial p_e}{\partial s} = eE_\perp$$

$$v_j\frac{\partial v_j}{\partial s} + \frac{1}{N_j m_j}\frac{\partial p_j}{\partial s} - \boldsymbol{g}\cdot\hat{\boldsymbol{s}} - \frac{eE_\parallel}{m_j} = -\sum_k v_{jk}(v_j - v_k). \tag{5-40}$$

Eliminating the electric field and introducing the local ion Mach number $M_j = v_j/c_j$, with $c_j = (2\mathcal{k}T_j/m_j)^{\frac{1}{2}}$ the ion sound speed, the equation for the ion plasma motion can be expressed by

$$\frac{1}{M_j}\frac{\partial M_j}{\partial s}(M_j^2 - 1) = \frac{\boldsymbol{g}\cdot\hat{\boldsymbol{s}}}{c_j^2} + \frac{1}{A}\frac{\partial A}{\partial s} - \frac{1}{c_j^2}\sum_k v_{jk}(v_j - v_k) - \frac{1}{F_j}\frac{\partial F_j}{\partial s}$$

$$- \frac{1}{2T_j}\frac{\partial T_j}{\partial s}(M_j^2 + 1) - \frac{1}{T_j}\frac{\partial T_e}{\partial s} - \frac{(T_e/T_i)}{N_e}\frac{\partial N_e}{\partial s} \tag{5-41}$$

$\star$ For a dipole field $ds = R_a \cos\varphi(4 - 3\cos^2\varphi)^{\frac{1}{2}}d\varphi$.

where A is the area of a magnetic flux tube $A \propto 1/|B|$ and

$$F_j = N_j(s) \cdot v_j(s) \cdot A(s) = \int_{s_0}^{s} (q_j - L_j) A \, ds + N_0 v_j A_0$$

is the ion escape flux, with q_j the ion production rate and L_j the chemical loss rate. Solutions of (5–41) can be found for appropriate boundary conditions. An example of such solutions for H^+ in a predominantly O^+ ionosphere is illustrated in Fig. 43. At high altitudes $M(H^+) \geq 0$ while at low altitudes, where chemical processes are important, $M(H^+) \to 0$.

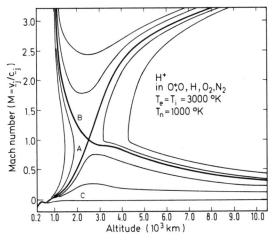

Fig. 43. Solutions for the Mach number $M(H^+)$ illustrating supersonic flow (polar wind)—curve A; subsonic outflow ("polar breeze")—curve B and diffusive equilibrium—curve C. (After Banks and Holzer [135])

The boundary plasma pressure $p(\infty)$ determines the appropriate solution. For outside plasma pressures $0 \leq p(\infty) < p_B(\infty)$, where $p_B(\infty)$ is the asymptotic pressure of the descending critical solution (B), curve A is the only correct solution, representing an *outward* flow of plasma which reaches supersonic velocity at the *critical point*, $M(H^+) = 1$.

Curve A applies ideally when $p(\infty) \to 0$ as $s \to \infty$. If $p(\infty) > p_B$ subsonic escape solutions ($M < 1$) apply, while for some $p(\infty)$, $M(H^+) \to 0$, diffusive equilibrium (curve C) applies.

Ion density profiles and H^+ escape fluxes [136] are shown in Fig. 44 for a particular model ionosphere. It can be seen that as the result of the upward flux of H^+, the scale height of H^+ becomes comparable to that of O^+ $(d(\ln N(H^+))/dz \cong d(\ln N(O^+))/dz)$ as discussed for diffusive flow.

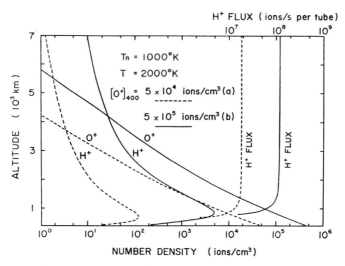

Fig. 44. Density distributions of H^+ and O^+ and escape flux of H^+ in the polar wind regime. (After Marubashi [136])

The asymptotic solution for the ion number density in the polar wind is given by

$$N_i \propto B[\ln(B)]^{-\frac{1}{2}} \propto R^{-3}.$$

The evaporative approach (ion-exosphere) can lead to supersonic flow velocities as the result of a polarization field obtained by requiring quasi-neutrality and the escape *fluxes* of ions and electrons to be equal [137] (rather than diffusive equilibrium, i.e., $v_e = v_i = 0$), or by the incorporation of a bulk flow *below* the ion-exobase [136]. While both, the modified evaporative and the hydrodynamic approach lead to supersonic velocities above the ion-exobase and to comparable escape fluxes, they both employ approximations which may not fully represent the true physical situation. In fact, the two approaches are complementary since the hydrodynamic approach is only appropriate in the collision-dominated region, whereas the kinetic (evaporative) approach can be applied only in the "collisionless" domain.★

★ For a detailed discussion see J. Lemaire and M. Scherer, Revs. Geophys. Space Phys. 11, 421—468 (1973).

Chapter VI

Models of Planetary Ionospheres

The distribution of charged particles (electrons and ions) in a planetary ionosphere is governed by the equation of continuity

$$\frac{\partial N}{\partial t} = q - L(N) - \mathbf{V} \cdot (N\,\boldsymbol{v}).$$ (6–1)

A steady state situation ($\partial N/\partial t = 0$) can often be assumed, except near sunrise and sunset or during other rapidly varying situations such as a solar eclipse.

VI.1. Equilibrium Models

There are two limiting cases to the continuity equation
1) *Chemical equilibrium* ($q = L$), when chemical processes are predominant and
2) *Diffusive equilibrium* ($\mathbf{V} \cdot (N\,\boldsymbol{v}) = 0$) when chemical processes can be neglected.

The applicability of either of these equilibrium models is determined by the appropriate time constants, i.e., the diffusion time $\tau_D \cong H_D^2/D_a$ and the chemical life time $\tau_C = N/L$. The smaller of the two time constants determines the appropriate equilibrium model as an approximation to the full continuity equation.

Thus, if $\tau_C \ll \tau_D$, chemical equilibrium prevails. The classical example of a chemical (photochemical) equilibrium model for a planetary ionosphere is the *Chapman layer*, which corresponds to

$$q = \alpha N^2$$

with q represented by the photoionization production function (2–11) and the chemical loss rate due to a recombination process (cf. 4–13).

The Chapman model distribution of plasma density is given by

$$N = N_m \exp\left\{ \frac{1}{2}\left[1 - \frac{z}{H} - \sec\chi \exp\left(-\frac{z}{H} \right) \right] \right\}$$

where   (6–2)

$$N_m = \left(\frac{q_m}{\alpha}\right)^{\frac{1}{2}} = \left(\frac{\eta_i \, \Phi_\infty \cos \chi}{e \, H \, \alpha}\right)^{\frac{1}{2}}.$$

The most important property of a Chapman layer is, that the maximum of electron (ion) density is identical to the height of the ion production rate maximum which is determined by the condition requiring that the optical depth $\tau(z) = \sigma_a n(z) H \sec \chi = 1$, occuring at an altitude (2–9)

$$h_m(\chi) \equiv h^*(\chi) = h_0^* + H \ln \sec \chi \, .$$

A Chapman layer will therefore be appropriate for a situation where $\tau = 1$ is satisfied at low enough altitudes, so that diffusion and chemical processes other than recombination are negligible, i.e., when the ionizable consituent corresponds to a molecular species (e.g., the terrestrial E layer). If on the other hand the ionizable constituent is an atomic species, and neutral molecular species are also present, then chemical loss processes involving molecular species (e.g., charge exchange reactions of the type $X^+ + YZ \rightarrow X Y^+ + Z$) will be much faster than radiative recombination in removing the atomic ions, although the final charge neutralization may occur as dissociative recombination of the secondary molecular ions. In this case the chemical equilibrium is given by

$$q = \beta(z) N \qquad\qquad\qquad (6–3)$$

which represents a height-dependent linear loss process, where the loss coefficient $\beta(z) = k n(YZ)$, leading to

$$N = \frac{q}{k n(YZ)}, \qquad\qquad\qquad (6–4)$$

i.e., N *increases* well beyond the level of maximum ion production, since $k n(YZ)$ *decreases* with altitude. At some altitude the further increase of N will be limited by diffusion, i.e., when $\tau_D \lesssim \tau_C$. A chemical equilibrium layer exhibiting this feature is often referred to as a *Bradbury layer*. (The terrestrial F_2 region below the F_2 peak is an example of such a distribution; this is illustrated in Fig. 46.) From the solution of the steady state continuity equation it can be shown that the maximum of the plasma density distribution will occur where $\tau_D \cong \tau_C$ while its value is given by the chemical equilibrium formula [138]

$$N_m \cong \frac{q(h_m)}{\beta(h_m)} \cong \frac{q(h_m)}{k n(YZ | h_m)}. \qquad\qquad (6–5)$$

Although the shape of the layer is controlled by chemical processes *and* diffusion, it can expressed empirically by a Chapman-like distribution of the type [139]

$$N = N_m \exp\left[\frac{1}{2}\left\{1 - \frac{\frac{z}{H}}{1-a\exp\left(\frac{az}{H}\right)} - \exp\left[\frac{-\frac{z}{H}}{1-a\exp\left(-\frac{az}{2H}\right)}\right]\right\}\right] \quad (6\text{--}6)$$

where the parameter $a=(H-H_0)/H$ represents the departure from a simple Chapman layer resulting from a variable scale height. At altitudes well above the ionization peak N_m, it reduces to $N \propto \exp[-z/2H]$ which (fortuitously) corresponds to a diffusive equilibrium distribution for $T_e = T_i = T_n$, i.e., $\mathscr{H} = 2H$. A Chapman layer and modified Chapman layers are illustrated in Fig. 45.

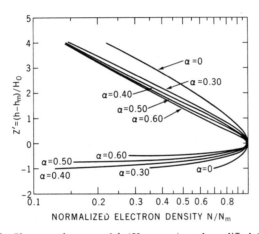

Fig. 45. Simple Chapman layer model (H=const) and modified Chapman layer (exponentially varying scale height). The parameter $a=(H-H_0)/H$ represents the degree of departure from the ideal Chapman layer. (After Chandra [139])

VI.2. Observables and Derived Parameters

From observations of the electron density distribution in a planetary ionosphere a number of physical characteristics can be inferred. If the ionization peak is at a relatively low altitude, for which the condition of unit optical depth for overhead sun can be satisfied, $\tau = 1$, the maximum density is given by

$$N_m = \left(\frac{\Phi_\infty \cos\chi}{e\,H\,\alpha}\right)^{\frac{1}{2}} \quad (6\text{--}7)$$

which is independent of the concentration of the number density of the ionizable constituent, but depends on the scale height of this (neutral) constituent as well as the ionizing flux outside the atmosphere and the recombination coefficient. The scale height can be determined from the observed decrease of electron density above the peak, according to

$$\frac{d(\ln N)}{dz} = -\frac{1}{H_N} \tag{6-8}$$

where $H_N = 2H$. With this information the consistency of a Chapman-layer assumption can be checked.

The number density of the ionizable constituent at the altitude of the ionization peak h_m can be determined from the relation

$$n_m(X) = \cos\chi(\sigma_a H)^{-1}. \tag{6-9}$$

It should be noted, that the observed scale height of a Chapman-layer allows the determination of the neutral gas temperature only; it does not provide any information regarding the temperature of the electrons and ions or the inference on the absence of thermal equilibrium.

For a "Bradbury layer", the concentration of the ionizable atomic constituent (X) as well as the molecular species (YZ) responsible for the chemical loss process can be inferred from a knowledge of N_m. The ionization maximum occurs at an altitude $h(N_m) > h(q_m)$ where $\tau_C \cong \tau_D$ (Fig. 46),

$$h_m \cong H_{eff} \ln\left(\frac{D_0}{H_{(X)}^2 \cdot k n_0(YZ)}\right) + h_0 \tag{6-10}$$

where $H_{eff} = \ell T/[m(X) + m(YZ)]g$ and h_0 is the reference level. In the presence of a plasma drift or wind vertical component velocity w, the height of the maximum will be changed by [140]

$$\Delta h_m \simeq \frac{w H^2}{D_a(h_m)} \simeq \frac{w}{\beta(h_m)}.$$

The height of the maximum h_m corresponds to the low-attenuation region, and $N_m = q_m/k n(YZ)_m$ can therefore expressed by

$$N_m = \frac{J_X n(X)_m}{k n(YZ)_m} \tag{6-11}$$

where $J_X = \langle \sigma_i \Phi_\infty \rangle$ is the ionization coefficient for species X. From the condition $\tau_C = \tau_D$, we obtain

$$\frac{\mathscr{B}}{n(X)H^2} = k n(YZ) \tag{6-12}$$

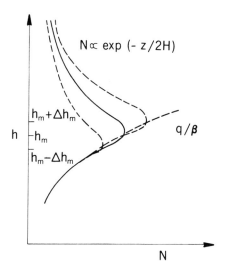

Fig. 46. Model of an F region, illustrating a "Bradbury layer" where the lower part is in chemical equilibrium with a linear loss law $N = q/\beta$, the peak is formed at h_m by plasma diffusion ($\tau_C = \tau_D$) and the distribution above the peak is controlled by plasma diffusion 1, i.e., $N \propto \exp(-z/2H)$. Also shown is the effect on h_m of the vertical component of horizontal winds or of plasma drift. (After Rishbeth)

if it is assumed that the ambipolar diffusion takes place through the ionizable constituent (X) and thus, $\mathscr{B}/n(X) \equiv D_a$. It then follows that

$$n(YZ|h_m) = \frac{1}{kH} \left(\frac{\mathscr{B} J_x}{N_m} \right)^{\frac{1}{2}} \tag{6-13}$$

and

$$n(X|h_m) = \frac{1}{H} \left(\frac{\mathscr{B}}{J_x} N_m \right)^{\frac{1}{2}}. \tag{6-14}$$

The scale height can be derived from the decay of the electron density well above the peak

$$-\left(\frac{1}{N} \frac{\partial N}{\partial z} \right)^{-1} \equiv H_N = \frac{\mathscr{k}(T_e + T_i)}{m_+ g}$$

assuming that the plasma temperature is constant. If this is not the case, the observed density scale height is modified by the temperature gradients (cf. 5–18).

It should be emphasized that the scale height of a diffusive equilibrium distribution is determined by the charged particle temperatures T_e and T_i and the mean ionic mass m_+, in contrast to that of a chemical equilibrium distribution, which depends on the temperature of the ionizable constituent, i. e., T_n. At altitudes well above the ionization peak, N_m, the plasma density distribution is generally well represented by a diffusive equilibrium distribution (cf. Chapter V)

$$N = N_0 \exp - \int\limits_0^z \frac{dz}{H} \equiv N_0 \exp - \int\limits_0^z \frac{m_+ g}{\mathscr{k}(T_e + T_i)} dz . \qquad (6\text{–}15)$$

If more than one ionic constituent is present, then the electron density distribution can be expressed by [141]

$$N = N_0 \left[\frac{\sum\limits_j N_{jo} \exp\left(-\dfrac{z}{H_j}\right)}{\sum\limits_j N_{jo}} \right]^{1+\varepsilon} \qquad (6\text{–}16)$$

where $H_j = \mathscr{k} T_i / m_j g$ and $\varepsilon = (T_e / T_i)$.

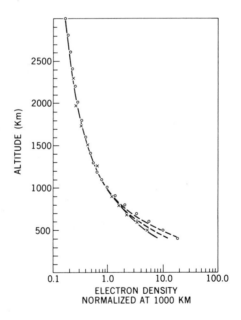

Fig. 47. Normalized electron density profiles in the topside ionosphere. Dashed lines: minimum and maximum densities for 10 topside sounder profiles over a midlatitude station; solid line and crosses: individual Alouette 2 profiles; circles: ternary ion-mixture, diffusive equilibrium model according to equation (6–17). (Courtesy of J. E. Jackson)

For a ternary ion mixture, and $T_e = T_i$, the plasma density distribution in diffusive equilibrium is given by

$$N = N_0 \left\{ \exp \frac{1}{2} \left[-\left(\frac{z}{H_1}\right) - \ln\left(1 + \eta_{21} \exp\left(\frac{z}{H_{12}}\right) + \eta_{31} \exp\left(\frac{z}{H_{13}}\right)\right) \right. \right.$$

$$\left. \left. + \ln(1 + \eta_{21} + \eta_{31}) \right] \right\} \qquad (6\text{--}17)$$

where $H_{ij} = \mathscr{k} T/(m_i - m_j)g$ and η_{ij} is the ratio N_{io}/N_{jo} at the reference level where the plasma density is N_0. Such a distribution is illustrated in Fig. 47.

Another parameter of an ionospheric layer which may be observable is the total or integrated content. For a simple Chapman layer one obtains upon integration in terms of the error function [142]

$$N_T = \int_0^\infty N_m \exp \frac{1}{2}\left(1 - \frac{z}{H} - e^{-\frac{z}{H}}\right) dz = N_m \cdot H \sqrt{2\pi e} = 4.13\, N_m \cdot H. \quad (6\text{--}18)$$

For a quasi-Chapman function as given in (6–6), the total content will be less than for the simple Chapman function, i. e., roughly reduced by the factor $(1-a)$.

The ratio of the electron content above the ionization peak to that below for a Chapman layer is given by

$$\frac{N_a}{N_b} = \frac{\mathrm{erf}(\frac{1}{2}\sqrt{2})}{1 - \mathrm{erf}(\frac{1}{2}\sqrt{2})} = 2.15. \qquad (6\text{--}19)$$

For a modified Chapman layer with a constant scale height gradient $dH/dz = \beta = \mathrm{const}$

$$N = N_m \left(\frac{1+\beta}{2}\right) [1 - \xi - e^{-\xi}] \qquad (6\text{--}20)$$

this ratio is greater than 2.15 and can be expressed in terms of the incomplete (Γ_k) and complete (Γ_∞) Gamma functions*, according to

$$\frac{N_a}{N_b} = \frac{\Gamma_{\frac{1+\beta}{2}}[\frac{1}{2}(1-\beta)]}{\Gamma_\infty[\frac{1}{2}(1-\beta)] - \Gamma_{\frac{1+\beta}{2}}[\frac{1}{2}(1-\beta)]}. \qquad (6\text{--}21)$$

$\star \ \Gamma_\infty(p+1) = \int_0^\infty x^p e^{-x} dx$ and $\Gamma_k(p+1) = \int_0^k x^p e^{-x} dx$.

VI.3. Ionospheric Regions

In analogy with the terrestrial ionosphere the following ionospheric regions may be identified (although not all of them may be present) in a planetary ionosphere in order of increasing altitude of their peak (Fig. 48).

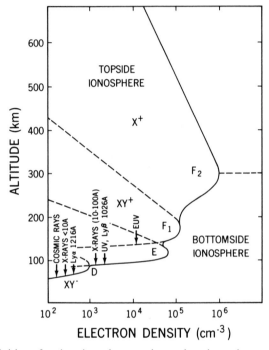

Fig. 48. Definition of regions in a planetary ionosphere in analogy to the terrestrial ionosphere. The principal ionization sources are also indicated

D-Region: This region represents the lowermost part of a planetary ionosphere where the ionization sources are galactic and solar cosmic rays and the most penetrating EUV lines, such as Lyman-α. The unit optical depth condition for this region requires a column content of the absorbing constituent of $\mathcal{N} \gtrsim 10^{20}$ cm^{-2}. A Chapman layer model is generally applicable, with molecular ions as principal constituent and dissociative recombination and three body-reactions being important chemical processes. Because of the relatively high number densities, ion clusters and negative ions resulting from electron attachment are present in this region.

E-Region: The ionization sources for this region are solar X-rays and EUV radiation, low energy ($\sim$ keV) energetic particles and to a lesser extent meteors. Unit optical depth for this region requires a total content of the absorbing constituent of $\mathcal{N} \simeq 10^{19}\, \mathrm{cm}^{-2}$. A Chapman layer model is applicable; molecular ions produced by the ionizing radiations are being lost by dissociative recombination. Because of the short chemical time constant, plasma transport processes are essentially negligible.

F_1-*Region:* The ionization source for this region is the least penetrating EUV radiation. Unit optical depth requires a total content of the absorbing constituent of $\mathcal{N} \simeq 10^{17}\, \mathrm{cm}^{-2}$. The F_1 region has an ion production peak; however, on Earth it is usually not noticeable as a distinct layer but only as a ledge in the plasma density distribution. Because of the presence of molecular ions, the chemistry is controlled by dissociative recombination. The F_1 region also conforms to a Chapman layer.

$F_2(F)$-*Region:* This region consists of atomic ions whose production maximum is in the F_1 region. The electron density increases beyond the ion production peak due to the presence of atomic ions which are lost by charge transfer processes with molecular neutral species, according to a linear loss law. This region represents a "Bradbury layer". The electron density peak (on Earth the F_2 peak represents the absolute maximum) occurs at an altitude where the chemical and diffusion time constants are approximately equal, i. e., where $\beta \cong D_a/H_D^2$. This condition requires a neutral density at that level of $n(h_m) \approx \tau_C \cdot 10^{19}/H_D^2$, corresponding to values $10^7 \lesssim n(h_m) \lesssim 10^{10}\, \mathrm{cm}^{-3}$. The F_2 region is controlled by chemical processes (charge transfer) *and* plasma transport processes. In the topside ionosphere, i. e., above the F_2 peak, plasma diffusion is the dominant process controlling the electron density distribution[*].

Ion-Exosphere: This term refers to a region where the plasma mean free path is of the order of the plasma scale height, allowing ions with sufficient kinetic energy to escape from the planetary force field when magnetic flux tubes are open or no magnetic field is present. The base of an ion-exosphere can be defined by the condition $N_c \lesssim 10^{-4}\,(m_i/m_H)g\,T_e$. An ion-exosphere may or may not be linked to a planetary ionosphere, i.e., ion-exospheric conditions may apply at the planetary surface. (Planetary bodies having an ion-exosphere without an ionosphere are the Moon and Mercury.)

---

[*] The terrestrial topside ionosphere where H^+ is the predominant constituent is also called the *protonosphere*.

VI.4. Realistic Models of Planetary Ionospheres

While equilibrium solutions of the continuity equation as discussed in section 1, may provide a first order description for a particular region of a planetary ionosphere, they represent only approximations to the real situation.

   The behavior of the ionospheric plasma is governed by the equations of continuity, momentum and heat transport. Since the ionosphere arises from the interaction of ionizing radiations with the neutral atmosphere, the appropriate equations of the electron, ion and neutral gases are coupled. Any change in the energy input not only affects the thermal structure, but also the composition and ionization of the atmosphere.

   For realistic conditions, it is therefore necessary to solve simultaneously the equations of continuity

$$\frac{\partial N_j}{\partial t} = q_j - L_j - \nabla \cdot (N_j v_j) \tag{6-22}$$

the momentum equations

$$m_j N_j \left(\frac{\partial v_j}{\partial t} + v_j \cdot \nabla v_j \right) = \nabla p_j + m_j N_j g - e N_j \left(E + \frac{1}{c} [v_j \times B] \right)$$
$$- \sum_k K_{ik}(v_j - v_k) \tag{6-23}$$

and the equations of heat transport

$$\rho_j c_v \frac{\partial T_j}{\partial t} - \nabla \cdot [K(T) \nabla T_j] = Q_j - L_j \tag{6-24}$$

for the charged and neutral species.

   The above equations have the general form

$$\frac{\partial \Psi_j}{\partial t} = A_j \left(z, t, \Psi_j, \frac{\partial \Psi_j}{\partial t}, \Psi_k \right) \frac{\partial^2 \Psi}{\partial z^2} + B_j(z, t, \Psi_j, \Psi_k) \frac{\partial \Psi_j}{\partial z} + C_j(z, t, \Psi_j, \Psi_k). \tag{6-25}$$

The appropriate boundary conditions are expressed by

$$\Psi_j(z_l, t) = f_j(t)$$

$$\left. \frac{\partial \Psi_j}{\partial z} \right|_{z_u, t} = \phi_j(t)$$

where z_l is the lower and z_u is the upper boundary level. Since no theoretical methods are available for an analytical solution of such a set of nonlinear differential equations, one has to resort to linearizing procedures. High speed computers are particularly suited for an approach involving partial linearization together with iterative steps. Such an approach has been employed in the solution of the steady state [143] and time-dependent continuity equations [144] for the terrestrial ionosphere, and for the ionosphere of Venus [95] with the additional upper boundary condition of pressure balance between the solar wind and ionospheric plasma. Although the approach outlined above is time consuming, it provides for the self-consistent solution of both neutral atmosphere and ionosphere properties. In this fashion no ad hoc models for the neutral atmosphere need to be invoked which may not be consistent with all aspects of ionospheric behavior (e. g., the ionizable neutral constituent may also be responsible for heat loss of the neutral and ionized species and thus control the thermal structure as well).

The use of this approach is well justified when some of the important input parameters are known, from which secondary inputs such as the wind field or thermal structure can be incorporated in a consistent manner. In addition, semi-empirical ionospheric models [145] can be developed which fit boundary conditions given by observations.

VI.5. The Extent of Planetary Ionospheres

According to our definition, a planetary ionosphere represents the thermal plasma (electrons and ions) produced by the interaction of ionizing radiations with the neutral atmosphere, which is controlled by the force field (gravity, magnetic field) of the planet. Thus, a planetary ionosphere may be thought to extend as far as the actions of these force fields allow the plasma to be confined in the planet's vicinity.

The plasma density distribution for a rotating planet can be expressed in terms of the potential (per unit mass) ϕ_g of the gravitational and inertial (centrifugal) force and the potential ϕ_E of the electrostatic polarization field which ensures charge neutrality, according to

$$N(R) = N_M \exp\left[-\frac{m_i \phi_g' + e \phi_E'}{\not{k} T_i}\right] \qquad (6\text{--}26)$$

where

$$\phi_g' = \phi_g(R) - \phi_g(R_M) \quad \text{and} \quad \phi_g(R) = -\frac{g_0 R_0^2}{R} - \frac{1}{2}\Omega^2 R^2$$

with g_0 the surface gravity and Ω the angular speed of rotation of the planet of radius R_0 and

$$e \phi_E' = -\frac{m_i T_e}{T_e + T_i}\phi_g'.$$

Accordingly we obtain

$$N(R) = N_M \exp\left[\frac{-m_i \phi_g'}{\mathcal{k}(T_e + T_i)}\right].$$

(6–27)

Since the force field is related to the potential by $F = \nabla\phi$, the plasma density has a minimum where $\nabla\phi = 0$. This condition defines a distance R_∞, according to

$$\frac{g_0 R_0^2}{R_\infty^2} = \Omega^2 R_\infty,$$

i.e., where the gravitational force balances the centrifugal force.

Thus, the maximum distance R_∞ (in the equatorial plane) where the plasma can corotate with the planet is given by*

$$R_\infty = \sqrt[3]{\frac{g_0}{\Omega^2 R_0}} \cdot R_0.$$

(6–28)

Table 29 lists this distance R_∞, for planets with an established ionosphere.

Table 29

Planet	R_∞ (in planet. radii)
Venus**	16 $R_\female$
Earth	6.6 $R_\oplus$
Mars	5.9 R_δ
Jupiter	2.2 $R_{2\!\!\!4}$

For a magnetic planet the ionospheric plasma is also confined by the magnetic flux tubes and the ionosphere may be considered to extend to the limit of the planetary field. This limit is determined by its interaction with the solar wind. The domain where the planetary magnetic field dominates is called the *magnetosphere* [146]. Its boundary, the *magnetopause* is determined by the balance between the solar wind streaming pressure p_{sw} and the magnetic field pressure according to

$$p_{sw} = \frac{B_b^2}{8\pi}$$

(6–29)

---

* Unless the plasma is confined by a corotating magnetic field, effective corotation will cease at much smaller distances, i.e., where ion-exospheric conditions ($N_c \lesssim 10^{-4} (m_i/m_H) g T_e$) prevail.

** Assuming atmospheric superrotation (4 day-rotation).

where B_b is the planetary magnetic field at the boundary and

$$p_{sw} = K N m v^2 \cos^2 \psi \qquad (6\text{–}30)$$

with $K \simeq 1$ the accommodation coefficient, and N, m and v, the number density, the mass, and the flow velocity of the solar wind ions (protons), respectively, and ψ the solar wind aspect angle. Applying this pressure balance and expressing B_b in terms of the surface field B_S, according to

$$B_b \simeq \tfrac{1}{2} B_S L_b^{-3}$$

where $L = R_b/R_0$; the location of the magnetopause is given by

$$L_b = \left[\frac{B_S^2 (1 + 3 \sin^2 \varphi_{ss})}{4 \pi N m v^2} \right]^{\frac{1}{6}}. \qquad (6\text{–}31)$$

Here φ_{ss} is the magnetic latitude at the subsolar point, accounting for the inclination of the dipole axis and the planet's orbit to the ecliptic, i.e., the seasons. Ahead of this magnetic obstacle a bow shock is formed at a distance from the magnetopause, $\Delta L \simeq 0.25 L_b$.

For Earth, $L_b \sim 11$, while $R_\infty = 6.6 R_0$, and one would thus expect that with a corotating magnetosphere the ionosphere would extend to about $6.6 R_0$. However, in fact, the terrestrial ionosphere terminates at the plasmapause ($\sim 4 R_\oplus$) [147], since convective motions resulting from the interaction of the solar wind with the magnetosphere limit true corotation to distances less than L_b or even R_∞ [148].

The configuration of the terrestrial magnetosphere resulting from the solar wind interaction with the geomagnetic field is illustrated in Fig. 41. In the antisolar direction the magnetosphere exhibits an open tail extending to great distances ($\gtrsim 60 R_\oplus$), issuing over the polar caps which enables the outflow of thermal plasma *(polar wind)*.

According to present understanding (which is far from complete) the *plasmapause* which is identified by an abrupt decrease in plasma density by one to two orders of magnitude [147], represents the boundary between convective motions due to an $E \times B$ drift resulting from the solar-wind induced dawn-to-dusk convection electric field

$$E_{\text{conv}} \simeq \frac{1}{c} V_A B_{ip} \sim \text{kV}/R_0 \qquad (6\text{–}32\,\text{a})$$

where V_A is the Alfvén velocity (cf. 7–32) in the solar wind and B_{ip} is the interplanetary magnetic field, and the corotating plasma due to the corotation electric field,

$$E_{\text{rot}} = -\frac{1}{c} (\Omega \times R) \times B. \qquad (6\text{–}32\,\text{b})$$

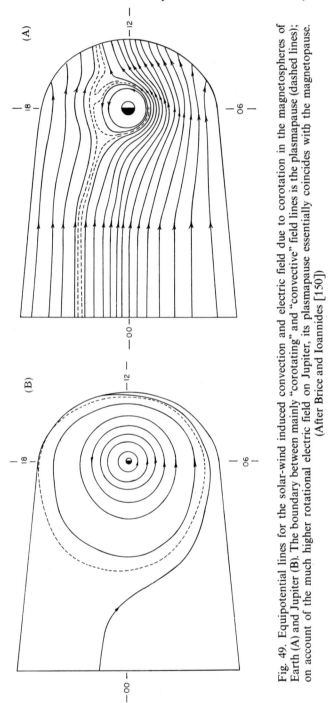

Fig. 49. Equipotential lines for the solar-wind induced convection and electric field due to corotation in the magnetospheres of Earth (A) and Jupiter (B). The boundary between mainly "corotating" and "convective" field lines is the plasmapause (dashed lines); on account of the much higher rotational electric field on Jupiter, its plasmapause essentially coincides with the magnetopause. (After Brice and Ioannides [150])

Figure 49 shows equipotential (stream) lines of these plasma motions with the plasmapause as a "stagnation" stream line [149, 150]. The actual configuration is even more complex. The plasmapause may be visualized as an asymmetric, irregular torus. The location of the plasmapause is strongly affected by changes in the solar wind interaction. An empirical correlation between the midnight plasmapause location and the planetary magnetic index has been found, according to $L_{pp} = 5.7 - 0.47\,K_p^*$, with K_p^* the maximum value in the preceeding 12 hours [147a].

In the case of Jupiter, the second magnetic planet (having a much stronger magnetic moment ($\mathscr{M} = R_0^3\,B_S$) than Earth, i.e., a surface field $B_{2\mathrm{1S}} \cong 10$ gauss), the ionosphere is thought to extend close to the location of the magnetopause, estimated to be at $\sim 50\,R_{2\mathrm{1}}$ [150]. Although the plasma density distribution will have a minimum at $R_\infty = 2.2\,R_{2\mathrm{1}}$, the Jovian magnetosphere is expected to corotate to much greater distances. The centrifugal force will accelerate ions beyond $L = 2.2$, where the density would increase again to a value which is eventually limited by recombination and flux tube interchange instabilities, (cf. Chapter VIII) and then decreasing according $N \propto L^{-4}$ [151].

For weakly or essentially non-magnetic planets such as Mars and Venus ($\mathscr{M} < 10^{-3}\,\mathscr{M}_\oplus$, where $\mathscr{M}_\oplus = 8 \times 10^{25}$ gauss cm^3) there may be no magnetosphere to shield the planetary ionosphere from a direct interaction with the solar wind*.

Since the solar wind with its frozen-in interplanetary magnetic field cannot penetrate another plasma, the solar wind is deflected around the ionosphere preventing its penetration to lower levels in the atmosphere where collision effects are dominant, or to the surface where it would be absorbed as is the case for the moon. Thus, the ionosphere represents an obstacle to the solar wind flow which leads to the formation of a bow shock upstream, similar to that where a magnetosphere is present [152, 152a].

The condition for pressure balance between the solar wind and ionospheric plasma, assuming that this pressure is much greater than the magnetic pressure on either side of this level ($p \gg B^2/8\pi$), can be expressed by

$$p_{sw} = K\rho_\infty v_{sw}^2 \cos^2\psi = N_i\mathscr{k}(T_e + T_i) \equiv p_{ion}$$

where $\rho_\infty = (Nm)_{sw}$ and v_{sw} are the solar wind values upstream of the bow shock and ψ is the angle between the flow direction of the undis-

---

* It now appears from the Mars 2 and 3 orbiter observations [151a], that Mars may possess a weak magnetosphere whose magnetopause at the subsolar point is at an altitude of ~ 1000 km, with $B_b \approx 20\gamma$. Thus, the Martian ionosphere may be protected from direct solar wind interaction [151b].

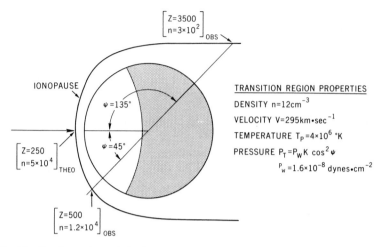

Fig. 50. Configuration of the ionopause of Venus resulting from the solar wind interaction with the topside ionosphere of Venus. Pressure balance between solar wind and ionospheric plasma is based on the Mariner 5 observations for the dayside occultation point ($\psi = 45°$); the shape of the ionopause is due to the $\cos^2 \psi$-dependence of the solar wind streaming pressure. In the antisolar direction there is a tail whose width is consistent with the nightside occultation data

turbed solar wind and the normal to the ionospheric boundary, which may be called the *ionopause*. The configuration of the ionopause due to the $\cos^2 \psi$ variation is illustrated schematically for Venus in Fig. 50. Also indicated are the locations of the two electron density profiles obtained from the Mariner 5 dual-frequency radio occultation experiment; the dayside profile exhibited a sharp drop in electron density at an altitude of ~ 500 km, which is interpreted as the ionopause, whereas the nightside profile extends to an altitude of ~ 3500 km (see Chapter IX). The thickness of the ionopause is probably of the order of the solar wind proton gyroradius [95, 250].

In addition to accumulation of magnetic fieldlines in front of the ionosphere forming an inpenetrable surface where a tangential discontinuity exists in the horizontal component of the interplanetary magnetic field and the solar wind velocity, an induced magnetosphere representing a magnetic barrier which provides for a pressure balance with the solar wind has been suggested as being responsible for the ionopause as observed for Venus. In either case a bow shock would be formed upstream as the result of the interaction of the supersonic solar wind with the planetary environment [152].

In connection with early data on the ionosphere of Mars where no sharp discontinuity or ionopause was observed, a different type of interaction has been suggested [153]. For this type of interaction which may be called soft in contrast to the previous hard interaction, it is assumed that the ions of the planetary ionosphere (photoions) are being incorporated into the solar wind flow at altitudes where a pressure balance cannot be satisfied (e. g., if the scale height is small due to a heavy constituent such as CO_2^+, with no light ions present). Unless the solar wind is to become subsonic, the addition of photoions should not exceed the solar wind mass flow rate, i.e., the mass loading should be limited to [154]

$$S^* = \frac{(\rho v)_{sw}}{(\gamma^2 - 1)} \tag{6-33}$$

where γ is the exponent in the law of adiabatic expansion ($\rho V^\gamma = \text{const}$). The mass loss rate due to solar wind "scavenging" is given by [154]

$$\frac{dM}{dt} = 2\pi\bar{\alpha} H R_b (\rho v)_{sw} \tag{6-34}$$

where $\bar{\alpha} \approx 0.34$ is the average loading fraction, H is the scale height of the neutral constituent and R_b is the planetocentric distance of the flow stream line.

For essentially magnetic field-free planets (Venus, Mars) mass loading would be substantial, so that it could provide a sufficient obstacle to the solar wind leading to the formation of a bow shock. More recently it has been suggested that the solar wind could decelerate to subsonic velocity as the result of these photoions, which are coupled with the solar wind by an $\boldsymbol{E} \times \boldsymbol{B}$ drift, without the formation of a bow shock since the photoions swept up by the solar wind may have a thermal velocity greater than the solar wind velocity [155, 155a].

Mass loading due to photoions in a planetary ionosphere (such as was originally proposed for Mars) could lead to a deceleration of the solar wind to a velocity of the order of the sound speed in the ionosphere ($v_{sw} \rightarrow v' \sim c_i$) and thus to a downward transport of ionospheric plasma with a velocity $v' = \text{const}$ [153]. This effect can be expressed by the continuity equation

$$v' \frac{dN}{dz} = q - L .$$

Since in the topside ionosphere, the time constant for the transport process $\tau_{v'} \simeq H/v'$ (for $v' \sim c_i$) is much smaller than the chemical time constant $\tau_C = 1/(\alpha N)$, the loss term may be neglected and the plasma density distribution, assuming the major ion to be a molecular species

$(X\,Y^+)$ is then given by

$$N(X\,Y^+|z) = \frac{H(X\,Y)}{v'}\, J_{XY}\, n(X\,Y|z)$$

where $q = J_{XY}\, n(X\,Y|z)$ above the ion production peak (i.e., in the low-attenuation region).

For a CO_2^+ (or O_2^+) ionosphere, such as Mars, the plasma density distribution then decreases according to

$$N(X\,Y^+|z) \propto \exp\left(\frac{-z}{H(X\,Y)}\right),$$

i.e., with the scale height of the ionizable constituent, whereas the photochemical equilibrium (Chapman) distribution at these altitudes corresponds to

$$N(X\,Y^+|z) \propto \exp\left(\frac{-z}{2H(X\,Y)}\right).$$

Thus, as the result of the downward plasma transport initiated by the solar wind interaction, the scale height of the plasma density distribution is by a factor of 2 smaller than that of a Chapman distribution (or a diffusive equilibrium distribution). This process has been invoked to explain the Mariner 4 observation of the Martian ionosphere. However, Mariner 6 and 7 observations do not seem to require this process to be operative, casting some doubt on its importance.

A recent interpretation of the magnetometer measurements onboard the Russian Mars orbiters 2 and 3 [151a] implies that Mars possesses an intrinsic dipole magnetic field with a moment $\mathcal{M}_\delta \simeq 2.4 \times 10^{22}$ gauss cm^3 corresponding to a surface field $B_S \simeq 60\,\gamma$. This interpretation is based on several crossings of the Martian "magnetopause" whose location at the sub-solar-point was inferred to occur at an altitude of ~ 990 km with a magnetic field strength $B_b \simeq 20\,\gamma$ at the boundary. It therefore appears that the Martian ionosphere should be protected from a *direct* interaction with the solar wind since the magnetosphere boundary pressure balancing the ram pressure of the solar wind is much greater than the ionospheric pressure, i. e.,

$$p_{sw} = \frac{B_b^2}{8\pi} \gg p_{ion}.$$

Thus, the situation for Mars is more similar to Earth than to Venus. However, in contrast to Earth whose ionosphere terminates where $E_{conv} \approx E_{rot}$, i. e., at the plasmapause, for Mars $E_{conv} \gg E_{rot}$ throughout the ionosphere, so that solar-wind induced convective motions can

penetrate much closer to the planet. In this case, the Martian ionosphere should extend to altitudes where the chemical time constant
becomes larger than the time constant for solar-wind induced convective
motions. This leads to a limiting value of ion density at the Martian
"plasmapause" or perhaps more appropriately called "chemopause", of

$$N \lesssim \frac{v_E}{\alpha_D R_\delta} \tag{6-35}$$

where $v_E \simeq c E_{conv}/B$ is the solar-wind induced convection velocity and
α_D is the dissociative recombination coefficient for the principal ion
(O_2^+/CO_2^+) and R_δ is the Martian radius. Using appropriate values for
the parameters in (6–35) one obtains $N \lesssim 5 \times 10^3$ cm^{-3} implying that
the Martian ionosphere extends to an altitude of ~ 300 km [151b].

Although our present understanding of the detailed processes responsible for the termination of planetary ionospheres is far from
complete, it seems certain that the solar-wind interaction determines
the extent of the ionospheres of both magnetic and non-magnetic planets.

Chapter VII

The Ionosphere as a Plasma

VII.1. General Plasma Properties

A plasma represents a collection of charged particles (electrons and ions) which interact by long range (Coulomb) forces, exhibiting a coherent behavior as the result of space charge effects [cf. 156].

The most fundamental length unit of a plasma is the *Debye length* λ_D which represents the distance beyond which the Coulomb field of an individual particle is no longer felt due to the shielding effect of a cloud of particles of opposite charge. For this reason λ_D is also called the Debye *shielding distance*. This means that in spite of long range Coulomb forces, charged particles in a plasma do not interact individually at distances greater than λ_D, so that for the plasma as a whole, effective Coulomb collision frequencies can be used. The Debye length is given by

$$\lambda_D = \left(\frac{k\,T}{4\pi e^2\,N} \right)^{\frac{1}{2}} \tag{7-1}$$

and numerically by

$$\lambda_D = 6.9 \left(\frac{T}{N} \right)^{\frac{1}{2}} \text{cm}$$

where T is in °K and N in cm^{-3}.

Because of space charge effects the electron and ion densities have to be averaged over dimensions much greater than λ_D so that quasi-neutrality $(N_e - \sum N_i \ll N_e)$ can be assumed.

The concept of a plasma also requires that the number of particles within a *Debye sphere*

$$\hat{N}_D = \frac{4\pi}{3} N \lambda_D^3 \gg 1. \tag{7-2}$$

The maintenance of electrons and ions in a plasma against recombination requires their kinetic energy to be greater than the potential energy. The distance where both are equal, the so called critical distance or *critical impact parameter*, is given by

$$r_c = \frac{e^2}{\ell T} \qquad (7\text{--}3)$$

while the *plasma (electron) mean free path* for Rutherford scattering (90° deflections) is given by

$$\lambda_e = (4\pi r_c^2 N)^{-1} \cong 4 \times 10^4 \frac{T_e^2}{N} \text{ cm} . \qquad (7\text{--}4)$$

The Debye length can then also be expressed by

$$\lambda_D^2 = \lambda_e r_c . \qquad (7\text{--}4a)$$

Generally, for a plasma the condition holds that

$$r_c \ll N^{-\frac{1}{3}} \ll \lambda_D \ll \lambda_e .$$

Another important (dimensionless) parameter is the *Coulomb parameter*

$$\Lambda = \frac{\lambda_D}{r_c} = \frac{\lambda_e}{\lambda_D} = 3\hat{N}_D \qquad (7\text{--}5)$$

which measures the degree to which plasma phenomena dominate over individual particle phenomena. This quantity often appears as $\ln \Lambda$, the *Coulomb logarithm*.

A planetary ionosphere represents a relatively weakly ionized plasma, i. e., where the ratio of electron-ion collision frequency to electron-neutral collision frequency in some regions is of the order of unity. Plasma transport coefficients applicable to the ionospheric plasma have been discussed in appropriate sections of this book (cf. Chapter V).

Another fundamental characteristic of a plasma is the plasma frequency (f_N), which represents the characteristic *eigen* frequency for electrostatic disturbances in the plasma.

The angular plasma frequency is given by

$$2\pi f_N \equiv \omega_N = \frac{v_l}{\lambda_D} = \left(\frac{4\pi e^2 N}{m} \right)^{\frac{1}{2}} \qquad (7\text{--}6)$$

where $v_l = (\ell T/m)^{\frac{1}{2}}$ is the *(longitudinal)* thermal velocity. Generally, the plasma frequency is defined in terms of the electron component, for which the numerical value is given by

$$f_N \cong (81\,N)^{\frac{1}{2}}$$

where f_N is in kHz and N in cm^{-3}.

The role of the Coulomb parameter in determining the relative importance of collective (plasma) phenomena versus individual particle processes can be seen from the ratio

$$\frac{v_{ei}}{\omega_N} \propto \frac{\ln \Lambda}{\Lambda},$$

i.e., when the Coulomb parameter is large, plasma phenomena dominate.

Typical values of the fundamental plasma parameters for planetary ionospheres are listed in Table 30.

Table 30

λ_e	10^2 m to 10^3 km
λ_D	1 mm to 10 cm
Λ	10^5 to 10^7
$\ln \Lambda$	12 to 16
f_N	100 kHz to 10 MHz

For many problems the ionospheric plasma can be treated as a single fluid, since electrons and ions are closely coupled with macroscopic properties derived from averages over the particle population. For a partially ionized plasma such as is the case for certain parts of an ionosphere, the neutral gas can be considered a separate component coupled to the plasma by appropriate collision and energy transfer processes.

If the effects of interactions between individual particles and waves in a plasma need to be considered, then a *plasma kinetic* approach has to be adopted. The description of the plasma components is then by means of the *collisionless Boltzmann* (or *Vlasov*) equation which represents an expansion in Λ^{-1} of the Liouville equation for a plasma, whereas the next higher order expansion which includes collision effects is given by the *Fokker-Planck* equation.

VII.2. The Ionosphere as a Dispersive Medium

A plasma such as a planetary ionosphere has dielectric properties which permit the propagation of electromagnetic (transverse) as well as electrostatic (longitudinal) waves [156a]. These properties are a function of wave frequency making the ionosphere a dispersive medium.

Electromagnetic Waves

From a Fourier analysis of Maxwell's equations, expressing the radiation field in terms of plane waves $\propto \exp[i(k \cdot r - \omega t)]$ where ω is the (angular) wave frequency and k is the propagation vector ($|k| = \omega |n|/c$, with n the refractive index) one obtains the wave equation [157]

$$k \times (k \times E) + \left(\frac{\omega}{c}\right)^2 (K) \cdot E = 0 \qquad (7\text{-}7)$$

where (K) is the dielectric tensor, which is defined through the displacement

$$D = (K) \cdot E = E + \left(\frac{4\pi i}{\omega}\right) j \qquad (7\text{-}8)$$

as the sum of the vacuum displacement and the plasma current $j = Nev$ $(i = \sqrt{-1})$. In the presence of a magnetic field B, the velocity v derived from the particle momentum equation (neglecting thermal motion) is a function of B, expressed by a gyro-frequency $\omega_B = eB/mc$ as the result of the Lorentz $v \times B$ force.

The dielectric tensor (K) for $B \| \hat{z}$ can then be expressed by

$$(K) \cdot E = \begin{pmatrix} S & -iD & 0 \\ iD & S & 0 \\ 0 & 0 & P \end{pmatrix} \begin{pmatrix} E_x \\ E_y \\ E_z \end{pmatrix} \qquad (7\text{-}9)$$

where

$$S = \tfrac{1}{2}(R + L), \qquad D = \tfrac{1}{2}(R - L)$$

$$R = 1 - \sum_k \frac{\Pi_k^2}{\omega^2} \left(\frac{\omega}{\omega + Z_k \Omega_k}\right)$$

$$L = 1 - \sum_k \frac{\Pi_k^2}{\omega^2} \left(\frac{\omega}{\omega - Z_k \Omega_k}\right); \qquad P = 1 - \sum_k \frac{\Pi_k^2}{\omega^2}$$

with

$$\Pi_k^2 = \frac{4\pi N e^2}{m_k}$$

and

$$\Omega_k = \left| \frac{Z_k e B}{m_k c} \right|$$

the generalized plasma and gyrofrequency respectively, for either electrons or ions $\left(Z_k = \pm 1 \begin{Bmatrix} \text{ions} \\ \text{electrons} \end{Bmatrix} \right)$.

By expressing the propagation vector k in terms of the dimensionless vector n which has the magnitude of the refractive index $|n| = c/|v_{ph}|$, where the phase velocity $v_{ph} = \omega/k$ we obtain

$$n \times (n \times E) + (K) \cdot E = 0 \qquad (7\text{-}10)$$

or

$$\begin{pmatrix} S - n^2 \cos^2\theta & -iD & n^2 \cos\theta \sin\theta \\ iD & S - n^2 & 0 \\ n^2 \cos\theta \sin\theta & 0 & P - n^2 \sin^2\theta \end{pmatrix} \begin{pmatrix} E_x \\ E_y \\ E_z \end{pmatrix} = 0$$

where θ is the angle between the wave vector k and the magnetic field B is assumed to lie in the z-direction.

The condition for a non-trivial solution requires the determinant of the matrix to be zero, giving the *dispersion relation* (or the equation of the wave normal surface, represented by the loci of the tip of the vector n^{-1}).

One common form [158] of the dispersion relation is given by

$$\tan^2\theta = -\frac{P(n^2 - R)(n^2 - L)}{(Sn^2 - RL)(n^2 - P)}. \qquad (7\text{--}11)$$

The polarization of the electromagnetic waves can be expressed by the ratio (from the second line of 7–10)

$$\frac{iE_x}{E_y} = \frac{n^2 - S}{D} \qquad (7\text{--}12)$$

where $iE_x/E_y = \pm 1$ corresponds to righthanded or lefthanded circular polarization, respectively.

In the above derivation of the dispersion relation it was tacitly assumed that thermal motions of the plasma are negligible, i.e., that a cold plasma approximation is valid, which holds as long as the (longitudinal) thermal velocity v_l is much less than the phase velocity of the wave

$$v_l^2 \ll v_{ph}^2 \equiv \frac{\omega^2}{k^2}$$

or   (7–13)

$$n^2 \ll \frac{c^2}{v_l^2} \equiv \frac{mc^2}{\mathscr{k}T}.$$

This condition is well satisfied for many ionospheric applications. Probably the best-known dispersion relation for a cold plasma is the *Appleton-Hartree* formula which has been widely used in ionospheric work. This formula (for electron modes ignoring collisions) is usually expressed in the form [159]

$$n^2 = 1 - \frac{X}{1 - \left(\frac{\frac{1}{2}Y_T^2}{1 - X}\right) \pm \left\{\frac{\frac{1}{4}Y_T^4}{(1 - X)^2} + Y_L^2\right\}^{\frac{1}{2}}} \qquad (7\text{--}14)$$

where $X=(\omega_N/\omega)^2$ and $Y=\omega_B/\omega$ with $Y_L=Y\cos\theta$ and $Y_T=Y\sin\theta$ are the so called *magnetoionic parameters*.* If collisions with the neutrals are included, the refractive index becomes complex, by adding a term $-iZ$ (where $Z=v/\omega$, with v the effective collision frequency) wherever unity appears in the A. H. formula. Collisional effects are usually most important for low radio frequencies at low altitudes in a planetary ionosphere. The absorption of radio waves due to collisions can be expressed by $E\propto\exp(\kappa r)$, where the absorption coefficient κ is related to the imaginary part of n. Two limiting cases of absorption are usually considered: *nondeviative absorption* which occurs when the refractive index $n\to1$, having a frequency dependence ω^{-2}, and *deviative absorption* which occurs for conditions where $n\to0$, whose frequency dependence departs significantly from the inverse square law [160].

Other extensions to the A. H. formula, such as an energy-dependent collision frequency as well as the effects due to the presence of ions have also been considered [159].

The presence of a magnetic field makes the ionospheric plasma anisotropic, so that two characteristic magneto-ionic modes appear, corresponding to the upper and lower sign in the denominator, which are called the ordinary (*o*) and extraordinary (*x*) mode, the latter having another branch called the *z*-mode. For practical purposes, approximations to the A. H. formula [159] can be used which are called *quasi-longitudinal* (*QL*) and *quasi-transverse* (*QT*) referring to the relative importance of terms containing the angle θ between propagation direction and the magnetic field.

These approximations are valid according to the Booker criterion [159]

$$Y_T^4\ll4\,Y_L^2(1-X)^2 \quad\ldots\quad QL$$
$$Y_T^4\gg4\,Y_L^2(1-X)^2 \quad\ldots\quad QT. \tag{7-15}$$

The refractive index for these conditions is given by

$$n^2\cong1-\frac{X}{1\pm Y_L} \qquad (QL)$$

$$\left.\begin{array}{l} n^2\cong1-\dfrac{X}{1+(1-X)\cot^2\theta} \\[3mm] n^2\cong1-\dfrac{X}{1-Y_T^2/(1-X)} \end{array}\right\}(QT). \tag{7-16}$$

* In the ionospheric literature f_H is generally used for the electron gyrofrequency (i. e., $\omega_B\equiv\omega_H$).

For the simple case of propagation along the magnetic field $(\theta=0)$

$$n^2 = 1 - \frac{X}{1 \pm Y}. \tag{7-17}$$

The condition $n^2 = 0$ gives the results*

$$
\begin{array}{lll}
X = 1 & f_o = f_N & \text{ordinary mode } (o) \\
X = 1 - Y & f_x^2 = f_N^2 + f_x f_B & \text{extraordinary mode } (x) \\
X = 1 + Y & f_z^2 = f_N^2 - f_z f_B & z \text{ mode}
\end{array}
$$

which are referred to as *cutoffs* $(n^2 = 0)$, whereas the condition $n^2 = \infty$ represents *resonances* which are associated with electrostatic plasma waves to be discussed below.

Similarly, for propagation perpendicular to the magnetic field $(\theta = \pi/2)$

$$n^2 = 1 - X$$

$$n^2 = 1 - \frac{X(1-X)}{1 - X - Y^2}. \tag{7-18}$$

These results can also be recovered from the general dispersion relation (7–11) noting that in terms of the magnetoionic parameters X and Y

$$P = 1 - X \qquad R = 1 - \frac{X}{(1-Y)} \qquad L = 1 - \frac{X}{(1+Y)}$$

$$S = 1 - \frac{X}{(1-Y^2)} \quad \text{and} \quad D = \frac{XY}{(1-Y^2)}.$$

All modes discussed above are electron modes, i.e., the terms in the dispersion relation containing ion properties have been neglected due to their smallness.

An important electron mode is the so-called *whistler mode*, corresponding to $n^2 = 1 - X/(1-Y_L)$ for $Y > 1$, i. e., at frequencies below the local electron gyrofrequency.

Since waves in the ionosphere travel with a group velocity

$$v_g = \frac{\partial \omega}{\partial \mathbf{k}} = \frac{\partial \omega}{\partial k_x} \hat{\mathbf{x}} + \frac{\partial \omega}{\partial k_y} \hat{\mathbf{y}} + \frac{\partial \omega}{\partial k_z} \hat{\mathbf{z}} \tag{7-19}$$

individual frequency components of a wave packet will have different travel times.

A whistler represents the frequency components of sferics (i.e., electromagnetic noise produced by lightning discharges) arriving at

---

* This also corresponds to total reflection at vertical incidence.

different times as the result of the dispersive properties of the ionosphere, while travelling along magnetic field lines. Since for the whistler mode $n^2 \gg 1$, k can be expressed by

$$k = \frac{\omega_N}{c}\left(\frac{\omega}{\omega_B \cos \theta}\right)^{\frac{1}{2}}$$

and one obtains for the group velocity

$$v_g \equiv \frac{\partial \omega}{\partial k} \propto \omega^{\frac{1}{2}},$$

i.e., a descending tone *(whistler)* [161]. Using the more appropriate form for the refractive index, $n^2 = R$, it can be shown (for $f_B = \text{const}$) that in addition to the falling tone, there is a rising component above the so-called *nose frequency* $(f_n = f_B/4)$* representing a nose whistler. Observations of these nose whistlers have provided a wealth of information about the structure of Earth's outer ionosphere, including the identification of the plasmapause. In addition to the (electron) whistlers, *ion whistlers* have also been observed on satellites. These arise from the coupling of the electron whistler mode into an ion cyclotron mode at a

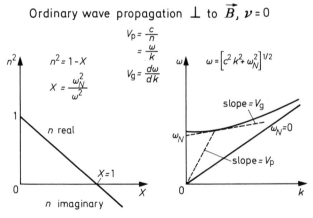

Fig. 51. Dispersion curves for ordinary mode propagation perpendicular to the magnetic field, ignoring collisions. Left panel shows a representation in terms of refractive index and magnetoionic parameter X as generally used in magnetoionic theory (Appleton-Hartree formula). Right panel shows general type of dispersion curve ω vs k which gives a direct representation of the group and phase velocity. (Courtesy of R. F. Benson)

---

* For propagation along a fieldline, $f_n \approx 0.38 \, f_{B_a}$, where the subscript a refers to the apex of the fieldline.

crossover $(D=0)$ i.e., where the refractive index for two wave modes is the same, which occurs between a pair of adjacent ion gyrofrequencies. These ion whistlers asymptotically approach the ion gyrofrequency.

Graphical presentations of the dispersion relations are made either in form of n^2 vs. X or ω vs. k as shown in Fig. 51, which also illustrates the relationship between propagation vector, phase velocity and group velocity, or in the form of the CMA (Clemmow-Mullaly-Allis) diagram, i.e., in $X-Y$, or $X-Y^2$ coordinates as shown in Fig. 52.

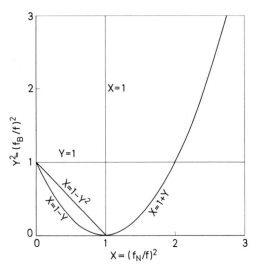

Fig. 52. CMA (Clemmow-Mullaly-Allis) diagram for electron modes, showing cutoffs and resonances as a function of the magnetoionic parameters X and Y^2

Plasma Waves

In addition to transverse electromagnetic waves, a plasma can also support longitudinal electrostatic oscillations or plasma waves as originally shown by Tonks and Langmuir [162]. This is simply illustrated for the case of no magnetic field when $K=n^2=1-X$. In this case the dispersion equation allows the solution

$$\frac{k^2 c^2}{\omega^2} = K$$

$$\boldsymbol{k} \cdot \boldsymbol{E} = 0$$

(7–20)

which represents *transverse electromagnetic waves,* but also

$$K = 0$$

$$\mathbf{k} \times \mathbf{E} = 0 \tag{7-21}$$

which corresponds to *longitudinal plasma waves* with a frequency $\omega = \omega_N$, i.e., the plasma frequency.

In the presence of a magnetic field, the plasma becomes anisotropic and the transverse and longitudinal modes cannot simply be separated into two dispersion relations, although the condition $\mathbf{k} \times \mathbf{E} \cong 0$ generally represents longitudinal waves while $\mathbf{k} \cdot \mathbf{E} \cong 0$ represents electromagnetic waves.

For longitudinal electrostatic waves, $\mathbf{E}$ is along the propagation vector $\mathbf{k}$ and the dispersion relation is given by

$$\mathbf{n} \cdot (\mathbf{K}) \cdot \mathbf{n} = 0 \tag{7-22}$$

which holds when $|n^2| \gg |K_{ij}|$ where the K_{ij} are the individual elements in the dielectric tensor which appear as the result of the anisotropy.

This implies that the refractive index for electrostatic waves is large and the phase velocity is slow.

If the thermal motions of the plasma are included, then, without a magnetic field, the dispersion relation for longitudinal waves $(K_L = 0)$ can be expressed by

$$\omega^2 \cong \omega_N^2 \left(1 + 3 \left(\frac{k v_l}{\omega_N} \right)^2 \right) \cong \omega_N^2 \left(1 + 3 \left(\frac{v_l}{v_{ph}} \right)^2 \right) \tag{7-23}$$

where the term in brackets represents the so-called *Bohm-Gross correction** to the Tonks-Langmuir formula [157]. In this form, the plasma waves are now propagating with a phase velocity

$$v_{ph} = \frac{\omega}{k} \tag{7-24}$$

and a group velocity

$$\frac{\partial \omega}{\partial k} \equiv v_g \cong \frac{3 k v_l^2}{\omega_N} . \tag{7-25}$$

From the complete dispersion relation it follows, that the propagation vector $\mathbf{k}$ has a real value only if ω is complex, i.e., a damping mechanism must be operative. This damping is not due to collisions, but rather due to resonances between the wave oscillations and particles whose velocity

---

* This represents the first term in an expansion applicable when $\langle v_e \rangle / v_{ph} \ll 1$; for an isotropic distribution $\langle v_e \rangle^2 = 3 v_l^2$.

equals the phase velocity of the wave $(v \cong v_{ph} < c)$, leading to an energy loss. This phenomenon is called *Landau damping*. It is unimportant as long as

$$k \lambda_{\mathrm{D}} = \frac{\omega}{\omega_N} \frac{v_l}{v_{ph}} \lesssim 0.2 \qquad (7\text{--}26)$$

but becomes highly effective when $k \lambda_{\mathrm{D}} \gtrsim 1^\star$ [162].

Another collisionless damping process is *cyclotron damping* [162]. Charged particles that feel oscillations at the cyclotron(gyro) frequency ω_B or its harmonics $n\, \omega_B$ can absorb energy from the wave. The resonant condition occurs where

$$\omega = -k_{\parallel} v_{\parallel} = n\, \omega_B, \qquad (7\text{--}27)$$

i.e., when the particles travel along the magnetic field with a velocity

$$v_{\parallel} = \frac{\omega - n\, \omega_B}{k_{\parallel}} \qquad (7\text{--}28)$$

where $k_{\parallel}$ is the propagation vector along B. This damping process involves particles with energies much higher than those responsible for Landau damping [162].

The intensity of plasma waves** is greatly enhanced by the presence of a non-maxwellian component (suprathermal particles). In planetary ionospheres photoelectrons represent such suprathermal particles which can lose energy by Coulomb collisions as well as by means of the *Čerenkov process*, leading to the generation of plasma waves (see Chapter III).

The condition for Čerenkov plasma wave generation is given by***

$$v_{ph} = \frac{\omega}{k} = v \cos \theta_c \qquad (v < c;\ n \gg 1) \qquad (7\text{--}29)$$

where v is the particle velocity and θ_c is the halfangle of the Čerenkov cone determined by

$$\frac{v}{c} n = \frac{1}{\cos \theta_c}.$$

Only particles whose velocity exceeds the rms thermal speed of the plasma $(v > (3)^{\frac{1}{2}} v_l)$ can excite waves by the Čerenkov mechanism, and as long as their velocity $v > 5 v_l$, Landau damping is not appreciable.

★ For $k \perp B$, Landau damping is absent, however, for intermediate directions to the magnetic field it may exceed the $k \parallel B$ (or no field) damping.

** Thermal background $(E_{\min})^2 \approx 4 \pi \mathscr{k} T / \lambda_{\mathrm{D}}^3$.

*** It should be noted that electromagnetic waves can also be generated directly by the Čerenkov mechanism, i.e., for particle velocities $v > 100\, v_{th}$, whereas plasma waves are generated by particles of velocity $v < 100\, v_{th}$ (cf. H. Oya, Radio Science 6, 1931, 1971).

These conditions are all met for ionospheric photoelectrons of energies $E > 3$ eV.

Plasma waves can also be coupled to electromagnetic wave modes when the phase velocities are nearly equal or converted into e.m. waves through density gradients or inhomogeneities. Plasma regimes where this is possible are illustrated in Fig. 53.

Plasma resonances occur where $n^2 = \infty$. From the dispersion relation (7–11) for a cold plasma one obtains the criterion for resonances [157]

$$\tan^2 \theta = -\frac{P}{S}. \tag{7–30}$$

For propagation parallel to the magnetic field $(k \times B = 0;$ i.e., $\theta = 0)$ resonances occur for $S = \frac{1}{2}(R + L) \to \pm\infty$ which correspond to the *electron cyclotron resonance*

$$\omega = \omega_B$$

and the *ion cyclotron resonance*

$$\omega = \Omega_B$$

and for $P = 0$ $(X = 1)$ representing the plasma resonance

$$\omega = \omega_N.$$

For propagation perpendicular to the magnetic field $(k \cdot B = 0;$ i.e., $\theta = \pi/2)$ resonances occur for $S = 0$ which are called *hybrid resonances*,

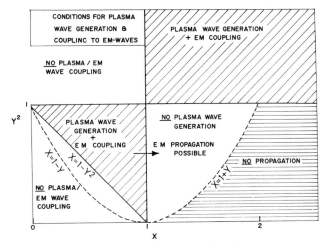

Fig. 53. CMA diagram depicting the conditions for plasma wave generation by the Čerenkov mechanism as well as coupling with and propagation of e. m. waves. (After Bauer and Stone [194])

with the *upper hybrid resonance* represented by

$$\omega_{UH}^2 = \omega_N^2 + \omega_B^2 \quad (\text{or } f_T = (f_N^2 + f_B^2)^{\frac{1}{2}}),$$

i.e., an electron mode corresponding to the magnetoionic condition $X = 1 - Y^2$, and the *lower hybrid resonance* representing an ion mode given by

$$\frac{1}{\omega_{LH}^2} = \frac{1}{\Omega_N^2 + \Omega_B^2} + \frac{1}{\Omega_B \omega_B}$$

where Ω_N is the 'ion' plasma frequency*.

These resonances are all *electrostatic* in nature even though they appear in the cold plasma approximation. Other types of plasma waves which appear only in a *warm plasma* are the *ion acoustic* and *ion cyclotron waves* occurring at frequencies below or near the ion gyrofrequency. The dispersion relation governing these waves is given by

$$n^2 \cong -\frac{P^*}{S \sin^2 \theta + P_i \cos^2 \theta} \tag{7-31}$$

where $P^* \equiv c^2/\lambda_D^2 \omega^2$ is the warm plasma correction factor and the subscript i refers to the ion properties.

For propagation $k \times B = 0$, the dispersion relation for the *ion acoustic wave* is given by

$$\frac{1}{\omega^2} = \frac{1}{k^2}\left(\frac{m_i}{\ell T_e}\right) + \frac{1}{\Omega_N^2}.$$

This holds for the condition $T_i \ll T_e$. If $T_i \simeq T_e$, the ion thermal velocity will become comparable to the ion acoustic wave phase velocity $(\ell T_e/m_i)^{\frac{1}{2}}$ and therefore suffer strong Landau damping.

The *ion cyclotron wave* appears for propagation angles $\theta \neq 0$ $(k \cdot B \approx 0)$ near the ion cyclotron frequency according to

$$\omega^2 \approx \Omega_B^2 + k^2\left(\frac{\ell T_e}{m_i}\right).$$

In a warm plasma electrostatic electron cyclotron waves are also possible which occur between cyclotron harmonics, approaching them asymptotically. They are called *Bernstein modes* $(n f_B < f_{Qn} < (n+1) f_B)$. For $\omega > \omega_{UH}$ $(f > f_T)$ and $k \cdot B = 0$, "resonances" exist at the point where $v_g \approx 0$; they are related to the electron gyrofrequency and plasma frequency according to [163]

$$f_{Qn} \simeq f_H\left[n + \frac{0.464}{n^2}\frac{f_N^2}{f_H^2}\right].$$

---

* We shall use capital Ω for the characteristic frequencies of the ions.

The cyclotron waves are strongly damped even for small departures from $\theta = \pi/2$ when $f \approx n f_H$.

Another type of "resonance" has been observed on ionograms obtained from the Alouette and ISIS satellites. It is the so-called *diffuse resonance* which is observed at the frequencies

$$f_{Dn} = f_{Qn+2} - 2 f_H$$

and weaker branches at the frequencies

$$f_{Dn1} = f_{Qn+1} - f_H.$$

It is not a true resonance in that the $v_g \approx 0$ condition is not satisfied and it appears to be the result of a plasma wave instability combined with nonlinear phenomena in the plasma [164].

Figure 54 shows plasma resonances (and cutoffs) observed on the Alouette and ISIS topside sounder satellites in the form of a f/f_H vs. f_N/f_H diagram.

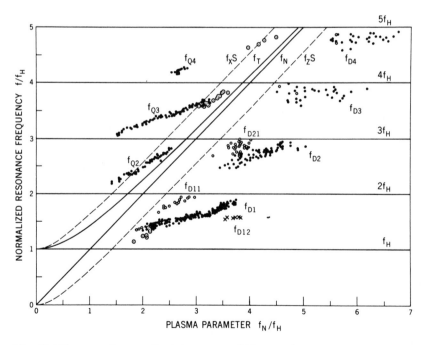

Fig. 54. Diagram of normalized frequency f/f_H vs. plasma parameter f_N/f_H, showing plasma resonances and cutoffs as observed on the topside sounder satellites [164]. (Courtesy of H. Oya)

Hydromagnetic Waves

These types of waves are very low frequency $(\omega < \Omega_B)$ waves which can propagate in a magneto-active plasma as the result of the combined action of hydrostatic and magnetic pressures [165].

The hydromagnetic mode for which the material velocity v is perpendicular to both the propagation vector and the external magnetic field $(k \cdot v = 0;\ B_0 \cdot v = 0)$ represents a *transverse* wave with a dispersion relation derived from small perturbation of the magnetic field $B_0 = B_0 \hat{z}$ according to

$$\omega = \frac{(B_0 \cdot k)}{(4 \pi \rho)^{\frac{1}{2}}}. \tag{7–32}$$

Since $B_0 \cdot k = B_0 k \cos \theta$, the phase velocity is

$$v_{ph} = \frac{\omega}{k} = V_A \cos \theta$$

where $V_A = B_0/(4 \pi \rho)^{\frac{1}{2}}$ is the *Alfvén velocity* with $\rho = N_i m_i + N_e m_e$ the mass density of the plasma. The group velocity for this transverse hydromagnetic wave *(Alfvén wave)* is given by

$$v_g = \frac{\partial \omega}{\partial k} = V_A \hat{z},$$

i.e., the disturbance travels *parallel* to the B field at the Alfvén velocity.

Other hydromagnetic modes can be obtained from the condition that v is coplanar with B_0 and k, i.e., these modes are neither purely transverse nor purely longitudinal. In this case the dispersion relation, given in terms of the phase velocity, corresponds to

$$v_{ph}^4 - (V_A^2 + c_s^2) v_{ph}^2 + c_s^2 V_A^2 \cos^2 \theta = 0 \tag{7–33}$$

where V_A is the Alfvén velocity and $c_s = (\gamma p_0/\rho_0)^{\frac{1}{2}}$ is the sound speed of the medium. From this equation (7–33), two modes are obtained: 1) the *modified Alfvén wave* which has a phase/group velocity $V_1 = V_A$ for propagation along the magnetic field $(k \times B = 0)$ and is in this case *transverse*, i.e., it degenerates into a simple Alfvén wave; whereas for propagation perpendicular to B_0 $(k \cdot B = 0)$, the wave is *longitudinal*, propagating with a phase/group velocity $V_1 = (V_A^2 + c_s^2)^{\frac{1}{2}}$; and 2) the *modified sound wave* which is longitudinal, propagating with the speed of sound c_s along B, but does not propagate perpendicular to the field when $c_s < V_A$. For $c_s > V_A$ the modified-sound wave propagates with a velocity $V_2 = (V_A^2 + c_s^2)^{\frac{1}{2}}$ across B, whereas in this case the modified

Alfvén wave does not propagate across the field. These mixed wave modes are also called *magnetoacoustic* or *magnetosonic waves*. The polar diagram of the phase velocities for the magnetosonic waves is illustrated in Fig. 55 for the case $c_s < V_A$ and for $c_s > V_A$ in parentheses.

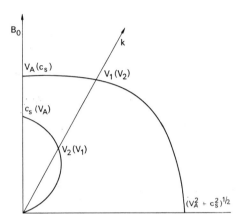

Fig. 55. Polar diagram showing the speed of hydromagnetic and modified hydromagnetic waves as function of angle between propagation vector and magnetic field. (After [165])

The Alfvén wave and the modified Alfvén wave can also be recovered from the cold plasma dispersion relation (7–11) for $\omega \ll \Omega_B$, yielding

$$n^2 \cos^2 \theta = 1 + \frac{\Omega_N^2}{\Omega_B^2}$$

$$n^2 = 1 + \frac{\Omega_N^2}{\Omega_B^2}.$$

(7–34)

Since $n^2 \gg 1$ and $(\Omega_N/\Omega_B)^2 = 4\pi\rho c^2/B_0^2$, one again obtains the phase and group velocities in terms of V_A.

The quasi-electromagnetic nature of hydromagnetic waves is also illustrated by the *hydromagnetic whistlers*, which result from the propagation of proton resonant magnetospheric emissions or *micropulsations* [166]. For $\omega \cong \Omega_B$, the cold plasma dispersion relation for $k \parallel B$, $n^2 = L$ is given by

$$n^2 \cong \frac{\Omega_N^2}{[\Omega_B(\Omega_B - \omega)]}$$

Table 31

Mode	Frequency	Condition	Type
Ordinary mode cutoff	ω_o	$\left.\begin{array}{l}P=0;\ X=1\end{array}\right\{$	electromagnetic
Plasma resonance	ω_N		electrostatic
Extraord. mode cutoff	ω_x ω_z	$R=0;\ X=1-Y$ $L=0;\ X=1+Y$	electromagnetic
Upper hybrid resonance	$\omega_{UH}(\omega_T)=(\omega_N^2+\omega_B^2)^{\frac{1}{2}}$	$S=0;\ X=1-Y^2$	electrostatic
Electron cyclotron res.	$\omega_B(\omega_H)$	$R\to\pm\infty;\ Y=1$	electrostatic
Ion cyclotron resonance	Ω_B	$L\to\pm\infty$	electrostatic
Lower hybrid resonance	ω_{LH} $(\omega_{LH}^{-2}=(\Omega_N^2+\Omega_B^2)^{-1}+(\Omega_B\omega_B)^{-1})$	$S=0$	electrostatic
Bernstein modes (warm plasma)	$n\,\omega_B\leq\omega\leq(n+1)\,\omega_B$	$\mathbf{k}\cdot\mathbf{B}=0$	electrostatic
Ion acoustic wave	$\omega\sim\Omega_N$	$\mathbf{k}\times\mathbf{B}=0$	electrostatic
Ion cyclotron wave	$\omega\sim\Omega_B$	$\mathbf{k}\cdot\dot{\mathbf{B}}=0$	warm plasma
Alfvén wave	$0<\omega\ll\Omega_B$	$\mathbf{k}\,\|\,\mathbf{B};\ v_g=V_A$	transverse hydromagnetic
Modified Alfvén wave (magnetosonic wave)	$0<\omega\ll\Omega_B$	$\left.\begin{array}{l}\mathbf{k}\,\|\,\mathbf{B};\ v_g=V_A \\ \mathbf{k}\perp\mathbf{B};\ v_g=(V_A^2+c_s^2)^{\frac{1}{2}}\end{array}\right\{$	mixed mode transverse/ longitudinal

and the group velocity

$$v_g \cong c\, \frac{\Omega_B}{\Omega_N} \frac{\left(1 - \dfrac{\omega}{\Omega_B}\right)^{\frac{3}{2}}}{\left(1 - \dfrac{\omega}{2\Omega_B}\right)} \cdot$$

For $\omega \to 0$, $v_g \to V_A$.

In contrast to the electron whistlers ($\omega < \omega_B/2$), hydromagnetic whistlers show a travel time which is increasing with frequency.

It has been suggested that the dissipation of hydromagnetic wave energy absorbed by the ionosphere, may represent a source of heating of the upper atmosphere [167]. Hydromagnetic power dissipation is most efficient at the 'higher' frequencies. This hydromagnetic heat source for a given frequency has a maximum at an altitude where

$$\omega_{hm}^2 \approx v_{in}\sigma_3 \left(\frac{V_{hm}}{c}\right)^2$$

with v_{in} the ion-neutral collision frequency, σ_3 the Cowling conductivity and the subscript hm referring to the particular hydromagnetic mode. Although for the terrestrial ionosphere, hydromagnetic heating is almost negligible compared to that by EUV, this may not be the case for some other planetary atmospheres.

The wave modes, cutoffs, and resonances that can occur in the ionospheric plasma are summarized in Table 31.

VII.3. Ionospheric Plasma Instabilities

Although there exist a large number of plasma instabilities we shall consider only the few which seem particularly pertinent to planetary ionospheres [168].

There are basically two types of plasma instabilities: 1) macroscopic and 2) microscopic, the first referring to unstable configurations in geometric space, the second to those in velocity space. The latter are of great importance in wave-particle interaction phenomena occuring primarily in the magnetosphere. Since the macroscopic instabilities refer to perturbations of the configuration of the plasma, they are especially important at the boundary of planetary ionospheres (ionopause, plasmapause).

The *gravitational (Rayleigh-Taylor) plasma instability* refers to the case where a dense fluid is floating on a lighter one separated by a boundary layer, with gravity acting on the system. Such a configuration is unstable to perturbations although surface tension in the boundary acts as a stabilizing force [169]. A similar situation exists in the interaction between the solar wind plasma and the ionosphere of a non-magnetic planet (e.g., Venus) in which case the boundary layer is represented by a magnetic barrier (due to the pile-up of interplanetary fieldlines or as the result of an "induced" magnetic field).

The condition for a stable configuration requires that the potential energy W be a minimum. Stability criteria can be developed by imposing a displacement and obtaining the change in W in terms of the variation δW.* In an equilibrium system $\delta W = 0$ to first order, a condition which is also called marginal stability. The system is truly stable when $\delta W > 0$; this condition corresponds to an increment in potential energy calculated to second order in the displacement [169].

For a magnetic boundary layer on top of a gravitationally supported plasma, a situation representative of the ionosphere boundary of Venus, the total potential energy of the system can be expressed by

$$W = \int \left(\frac{B^2}{8\pi} + \rho_i \phi_g \right) d\mathcal{V}$$

where $B^2/8\pi$ is the energy density of the boundary layer magnetic field which in turn is balanced by the solar wind streaming pressure p_{sw} (see 6–30), ρ_i is the mass density of the ionospheric plasma, ϕ_g is the gravitational potential $(\nabla \phi_g = -g)$ and $d\mathcal{V}$ is a volume element. The respective changes in gravitational potential and magnetic energy resulting from a displacement δR [169] are given in this particular case by

$$\delta W_g = \int \rho_i g \, \delta R \, d\mathcal{V}$$

and

$$\delta W_B = - \int \frac{B^2}{4\pi R_b} \delta R \, d\mathcal{V}$$

where R_b is the planetocentric distance of the ionopause. Thus, the total change in potential energy of the system can be expressed by

$$\delta W = \int \left[\rho_i g - \frac{B^2}{4\pi R_b} \right] \delta R \, d\mathcal{V}.$$

Accordingly, $\delta W > 0$ when $\rho_i g > B^2 / 4\pi R_b$.

---

* An alternate way of describing plasma instabilities is in terms of a wave dispersion relation, since perturbations of a stable equilibrium state result in harmonic oscillations (waves). Such an approach is particularly useful for determining the growth rate of the instability [168].

For a stable boundary we thus obtain the condition $(\delta W > 0)$

$$\rho_i g \equiv N_i m_i g > \frac{B^2}{4\pi R_b} \equiv \frac{2p_{sw}}{R_b}. \qquad (7\text{-}35)$$

Using the appropriate values for the Venus ionopause based on the Mariner 5 observations, a requirement for a stable ionosphere boundary is that $m_i > m(H^+)$, i. e., the presence of ions heavier than protons at the altitude of the ionopause [170].

For an ionospheric plasma contained by a dipole magnetic field (Earth, Jupiter), the charged particles experience an outward force in the curved flux tubes (especially where the centrifugal force due to planetary rotation becomes predominant), i. e., in the direction of decreasing particle density. This situation is again similar to the classical Rayleigh-Taylor instability*. Since the plasma is confined by magnetic flux tubes, the instability tends to interchange flux tubes. This type of instability is therefore also known as *interchange instability* [171].

For *adiabatic (fast)* interchange of flux tubes the condition for marginal stability $(\delta W = 0)$ can be expressed by [172]

$$\frac{d}{dR}\left[p\mathscr{V}^{\gamma}\right] = 0$$

where $p = N \mathscr{k} T_p$ is the plasma pressure (energy density) with $T_p = T_e + T_i$ the effective plasma temperature; $\mathscr{V}$ is the volume of a flux tube $(\mathscr{V} \propto R^4)$ and γ is the ratio of specific heats $(\frac{5}{8} < \gamma < \frac{7}{4})$, depending on the pitch angle distribution, with $\gamma = \frac{5}{3}$ a representative value.

From $p\mathscr{V}^{\gamma} = \text{const}$, we obtain

$$p \equiv N \mathscr{k} T_p \propto R^{-\frac{20}{3}}. \qquad (7\text{-}36\,a)$$

Thus, as long as the energy density of the ionospheric plasma decreases more slowly than $\propto R^{-7}$, (e. g., a plasma density distribution $N \propto R^{-(3.5 \pm 0.5)}$ as is found for the "collisionless" regime of the outer ionosphere, i. e., inside the Earth's plasmasphere), such a distribution should be stable against interchange (even if $T_p = \text{const}$). At the plasmapause, however, the plasma density decreases much faster and this configuration, as illustrated in Fig. 56, would be unstable unless the effective plasma temperature T_p increased significantly in the same region in order to satisfy the condition $p \propto R^{-7}$. This condition seems to be met

---

* Whereas the Rayleigh-Taylor instability refers to a hydrostatic boundary, the *Kelvin-Helmholtz instability* arises at a hydrodynamic boundary, i. e., when two incompressible fluids are in relative motion to each other.

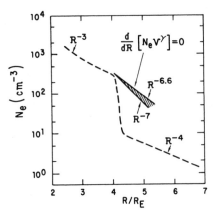

Fig. 56. Electron density profile in the equatorial plane of the plasmasphere from whistler observations and conditions for stability against plasma interchange. (After F. Scarf [172])

for the terrestrial plasmapause, where T_p is found to increase appreciably [172].

For *isothermal (slow) interchange* ("thermal convection") of flux tubes the marginal stability criterion is given by

$$p\mathscr{V} = \text{const}$$

and thus,

$$p \equiv N \mathscr{k} T_p \propto R^{-4} \qquad (7\text{--}36\,\text{b})$$

represents a plasma energy density distribution which is stable against slow (isothermal) interchange. This condition is also satisfied everywhere in the terrestrial plasmasphere, except near the plasmapause for $T_p = \text{const}$.

When the stability criterion is not satisfied, the amplitude a of perturbations will grow with time according to $a \propto \exp[(g'/L)^{\frac{1}{2}} t]$, where g' is the effective gravitational force per unit mass causing the instability (gravitational and/or centrifugal force) and L is the scale length of the density gradient; the *growth rate* (sec^{-1}) for the instability is given by

$$\gamma = \left(\frac{g'}{L}\right)^{\frac{1}{2}}. \qquad (7\text{--}37)$$

When the plasma density distribution follows $N \propto R^{-n}$, (where n is greater than the exponent required for marginal stability), the scale length is given by $L \sim R/n$. Instabilities induced by the centrifugal force $(g' = R\Omega^2)$ would then grow with a rate $\gamma = \sqrt{n}\,\Omega$; this process could be of importance in the Jovian ionosphere on account of the large angular rotation rate of Jupiter [151].

A plasma micro-instability known as *two-stream instability* can also be of importance in planetary ionospheres [168]. This instability occurs when two particle streams are flowing through each other. This is the case for ions in the polar wind or when the electrons and ions have different drift velocities in an electric field due to their difference in collision frequency with the neutrals and their gyrofrequencies. The latter condition applies to the dynamo regions.

The two-stream instability refers to the growth of longitudinal plasma ion waves which occurs when the relative velocity between the two streams (in the ideal case of a collisionless plasma without a magnetic field) is greater than the thermal velocity. Landau damping (important when the particle and wave velocities are comparable) has a stabilizing effect. The two-stream instability leads to the existence of plasma irregularities which travel with the ion acoustic velocity $(c_i = (2 \ell T / m_i)^{\frac{1}{2}})$.

For a partially ionized plasma, such as a planetary ionosphere, the $E \times B$ drift in the (dynamo) electrojet region leads to a two-stream instability even when the drift velocity of the electrons is $\sim v_{th}^{(i)} \sim 0.01 \, v_{th}^{(e)}$, for $T_i \sim T_e$. The velocity threshold for instability is lower than for the no field case (or propagation $\| B$), since Landau damping is absent for propagation of the ion plasma wave perpendicular to the magnetic field [173]. The gradient drift or $E \times B$ instability appears to be responsible for ionospheric irregularities associated with both sporadic E and spread F in the terrestrial ionosphere [173a, 173b].

Chapter VIII

Experimental Techniques

The fact that the ionospheric plasma is a dispersive medium (cf. VII.2) affecting the propagation of radio waves has made radio methods the first, and for some time the only, experimental technique for studying the terrestrial ionosphere [174]. The same holds true for the ionospheres of other planets since radio techniques allow one to obtain information on ionospheric properties from the transmission and reflection characteristics of radio waves received at a distant observation point, i. e., they represent *remote sensing techniques*. In contrast to these indirect measurement techniques, *direct* or *in situ* observation techniques require the instrument to be located within the ionospheric plasma, i. e., they are feasible only on spacecraft operating in the ionosphere.

VIII.1. Remote Sensing Radio Techniques

Dispersive Doppler and Faraday Effect

The path length of a radio wave traversing a dispersive medium such as a planetary ionosphere is given by

$$P = \int_{r_1}^{r_2} n \, dr \qquad (8\text{--}1)$$

where n is the refractive index, whereas the free space path is represented by

$$R = \int_{r_1}^{r_2} dr. \qquad (8\text{--}2)$$

The change in the phase path for a radio wave due to the presence of the ionosphere is thus

$$\Delta P = R - P \qquad (8\text{--}3)$$

and since $n < 1$, $P < R$, i. e., the phase path is reduced compared to the free space path length.

This equation can be written for the two magnetoionic modes, (the ordinary (o) and the extraordinary (x)) which can be separated by virtue of their different polarization, noting that the two phase paths are also different.

For a spacecraft moving in the ionospheric plasma, the *Doppler shift* ($f_D \equiv f' - f$, with f' the observed Doppler-shifted frequency transmitted from the spacecraft at frequency f) can be expressed by

$$f_D = -\frac{1}{\lambda}\frac{dP}{dt} \equiv -\frac{f}{c}\frac{dP}{dt} \qquad (8\text{-}4)$$

or by splitting it into the free space and dispersive component according to

$$f_D = -\frac{f}{c}\frac{dR}{dt} + \frac{f}{c}\frac{d\Delta P}{dt} = f_{Do} + \Delta f_D \qquad (8\text{-}5)$$

where f_{Do} is the free-space Doppler-shift and Δf_D the one due to the ionosphere. The quantity $dR/dt = v_r$ is the radial velocity component in the direction of the observer and $d(\Delta P)/dt$ can be expressed in terms of the refractive index n according to

$$\frac{d(\Delta P)}{dt} = n_s\frac{dr_s}{dt} + \int_{r_0}^{r_s}\frac{dn}{dt}dr \qquad (8\text{-}6)$$

where the subscript s refers to the location of the spacecraft and o to that of the observer, and the integral represents the time variation of n along the raypath. The dispersive Doppler effect was the first radio method to be employed on sounding rockets for the measurement of ionospheric electron densities. For increased accuracy, as well as reference purposes, harmonically related frequencies f and mf are used. The differential Doppler-shift for these frequencies provides a direct measure of the dispersive properties of the ionosphere, from which the electron density can be derived. The free-space Doppler shift can be removed by combining (out-of-phase), upon transmission through the ionosphere, the higher frequency mf with the lower frequency f after it is multiplied by the harmonic factor m, and the dispersive Doppler-shift is then given by

$$\mathscr{D}_{o,x} \equiv (mf)_D - mf_D = \frac{mf}{c}(n_{o,x}^{(mf)} - n_{o,x}^{(f)})\dot{r}_s + \int_{r_0}^{r_s}\frac{d}{dt}(n_{o,x}^{(mf)} - n_{o,x}^{(f)})dr \qquad (8\text{-}7)$$

where $n_{o,x} = \mathscr{F}(N_e)$ is the refractive index for the particular mode which is related to the electron density N_e through the Appleton-Hartree dispersion relation (cf. VII.2).

For frequencies $f \gg f_N$, f_B, the QL approximation is usually valid (except for $\mathbf{k} \perp \mathbf{B}$). In this case the two magnetoionic modes are circularly polarized with opposite senses of rotation. This fact leads to a rotation of the plane of polarization, the so-called *Faraday rotation* according to

$$\Omega_F = \frac{\omega}{2c}(\Delta P_o - \Delta P_x) \cong \frac{\omega}{2c}\int_{r_0}^{r_s}(n_o - n_x)\,dr \qquad (8\text{–}8)$$

where it is assumed that the distance between receiver and transmitter $(\overline{r_o r_s})$ remains fixed.

In the QL approximation $n_{o,x} \cong 1 - \frac{1}{2}X\,(1 \mp Y_L)$ and thus

$$n_o - n_x \simeq \frac{\omega_N^2\,\omega_B\cos\theta}{\omega^3}.$$

Hence, the Faraday rotation (in radians) between r_s and r_o is given by

$$\Omega_F = \frac{e^3}{2\pi m_e^2\,c^2\,f^2}\int_{r_0}^{r_s} N\,B\cos\theta\,dr \equiv \frac{K}{f^2}\,\overline{B\cos\theta}\int_{r_0}^{r_s} N\,dr \qquad (8\text{–}9)$$

where $K = 3.36 \times 10^8$ when f is given in MHz, B in gauss and N in cm^{-3}, for one-way propagation.

The Faraday rotation is thus a measure of the total electron *content* between transmitter and receiver. For lower frequencies, higher-order approximation of the dispersion relation have to be employed; consideration must also be given to refractive effects especially at low angles of elevation. The Faraday rotation technique has been extensively used, first with radio waves reflected from the moon and later with earth satellites; it is especially well-suited for geostationary satellites, providing the time variation of the total electron content of the terrestrial ionosphere [175].

When the pathlength changes as the result of the spacecraft's motion, the polarization (Faraday) fading on a linearly polarized receiving antenna is given by

$$\dot{\Omega}_F \cong \frac{K}{f^2}\,\overline{B}\left\{\frac{\partial}{\partial t}(\cos\theta)\int_{r_0}^{r_s} N\,dr + \cos\theta\,N_s\frac{dr_s}{dt}\right\}. \qquad (8\text{–}10)$$

Radio Occultation Technique

This technique utilizes transmitters or beacons on spacecraft, (usually those used for telemetry and tracking having frequencies in the range

$\sim 10^2$ to $\sim 10^3$ MHz); for US planetary probes the S-band (2200 MHz) system is used*, although, specially designed lower frequency transmitters have also been employed [176]. As the spacecraft in planetary fly-by trajectory or planetary orbit is *occulted* by the planet, the grazing ray path through the planetary ionosphere causes the observed Doppler-shift of the radio signals transmitted from the spacecraft to be different from the calculated (free space) Doppler-shift based on the spacecraft trajectory. The so-called Doppler residuals, Δf_D are the result of the dispersive properties of the medium. For a spacecraft receding from the observer, the residuals Δf_D caused by the ionosphere are $\Delta f < 0$, whereas those caused by the neutral atmosphere (troposphere) are $\Delta f > 0$. Since both, the ionosphere and the neutral atmosphere can be investigated by the occultation technique, the Doppler residuals are often expressed in terms of *refractivity* $\tilde{N} = (n-1) \cdot 10^6$. For the neutral atmosphere the refractivity $(\tilde{N} > 0)$ is a function of pressure, temperature and composition, whereas for the ionosphere $(\tilde{N} < 0)$ it is a function of the electron density N_e. Typical Doppler residuals obtained by the S-band radio occultation experiment [177] are shown in Fig. 57.

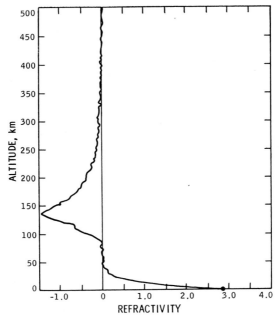

Fig. 57. Doppler residuals from S-band occultation experiment. Negative residuals represent the ionosphere, positive residuals the neutral atmosphere. Data are from Mariner Mars occultation experiment. (After [235])

---

* In the future both S- and X-band (8.8 GHz) systems will be used.

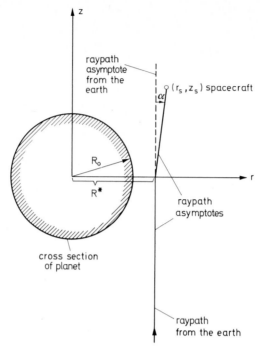

Fig. 58. Geometry depicting radio occultation experiment for the investigation
of a planetary ionosphere

It should be noted, that correction due to the dispersive effects of the interplanetary medium and the terrestrial ionosphere need to be made before the Doppler residuals are truly representative of the planetary environment. The phase path of the radio waves is also affected by refraction. This is illustrated in Fig. 58 which shows the geometry of the radio occultation experiment. For small angles of refraction, this angle is given by

$$\alpha = \frac{\Delta f_D c}{f} \left(\frac{dr_s}{dt} \right)^{-1}.$$

The phase path as a function of time can be expressed by

$$P(t) = -\frac{c}{f} \int_{t_1}^{t_2} \Delta f_D \, dt \qquad (8\text{--}11)$$

for imersion, i. e., entering into occultation, whereas for emersion, the opposite sign applies.

The observed phase change P represents the superposition of two effects due to the medium; 1) the change introduced by the variation in refractivity along the ray path, P, and 2) the phase path increase due to refraction $\Delta P_{ref} = \alpha^2 z_s/2$, so that

$$P_1 = P - \Delta P_{ref}$$

$$P_1 = 10^{-6} \int_{-\infty}^{+\infty} \tilde{N}(r)\,dz$$

where $\tilde{N}$ is the refractivity which for frequencies $f \gg 100$ MHz is related to the electron density by $\tilde{N} = -40.3\,N_e\,(cm^{-3})/f^2\,(MHz)$. Thus, the phase change expressed by the integral of the Doppler residuals represents a measure of the electron content along the ray path. By virtue of the Chapman function $Ch(\chi, R^*)$ this information can be directly related (for a spherically stratified ionosphere) to the electron content in the vertical direction over the point of closest approach of the ray path, $(R^* = \sqrt{r^2 - z^2})$. The total content along the grazing path through the planetary ionosphere is thus related to the vertical total content by

$$P_1 \propto \int_{-\infty}^{+\infty} N_e(r)\,dz = 2N_0 H \cdot Ch\left(\frac{\pi}{2}, R^*\right) = 2N_0 H \left(\frac{\pi}{2}\,\frac{R^*}{H}\right)^{\frac{1}{2}}$$

$$= N_0(R^*)(2\pi H R^*)^{\frac{1}{2}}.$$

(8–12)

The electron density at the point of closest approach $N_0(R^*)$ can then be determined together with information on the scale height H.* The conversion of P_1 to an altitude profile of N_e requires an inversion based on the Abel integral transform [176]**.

Generally, simplifying assumptions such as spherical symmetry are made which may not always be realistic. These assumptions can lead to errors in the electron density and particularly in the scale height; these errors scale according to H/R_0, i.e., the ratio of scale height to planetary radius. Thus, these errors are most serious for Mercury and least for Jupiter, with Mars and Venus representing intermediate cases, and they are especially serious for satellite orbits out of the ecliptic [178].

An improvement in the sensitivity of the occultation experiment for ionospheric measurements can be gained by employing two harmoni-

---

* In the presence of a scale height gradient β, the result is modified by the factor $\Gamma_\infty((1/\beta) - (\tfrac{1}{2}))/\beta^{\frac{1}{2}}\,\Gamma_\infty(1/\beta)$.

** For a detailed discussion see Proc. Workshop on Mathematics of Profile Inversion, L. Colin, ed., NASA TMX-62, 150, 1972.

cally related frequencies f and mf. In this case the total content is given by [176]

$$\int_{-\infty}^{+\infty} N\,dz \propto \frac{m^2 f^2 [P_1(\rho, mf) - P_1(\rho, f)]}{m^2 - 1}. \qquad (8\text{–}13)$$

Such a dual frequency occultation experiment was performed with the Mariner 5 spacecraft, employing frequencies of 49.8 MHz and 423.3 MHz, providing a measurement sensitivity in electron density of $\sim 10^2$ cm^{-3}. The minimum usable frequency for the ionospheric occultation experiment depends on the distance of the spacecraft from the planet and the maximum plasma frequency according to

$$f_{\min} \propto (z_s/H)^{\frac{1}{2}} f_N(\max).$$

Ionospheric Sounding

Sounding of the ionosphere by radio waves represents the oldest experimental technique for investigating the terrestrial ionosphere [174]. An *ionospheric sounder (ionosonde)* is basically a variable frequency radar which measures the time delay between the transmission of an rf pulse and its echo from a reflecting level in the ionosphere. Since the use of this technique on satellites, a distinction is made between *bottomside sounding*, i.e., from the ground and *topside sounding*, i.e., from a satellite, since for the first only the ionosphere below the F_2 peak (the bottomside ionosphere) is accessible, whereas with the latter only the topside ionosphere can be observed [179].

The time delay τ between the transmitted pulse and the echo defines a *virtual range P'* according to

$$P' = \tfrac{1}{2} c \tau \qquad (8\text{–}14)$$

where c is the velocity of light in vacuo. This path length is not the true one since in the ionosphere the signal velocity is represented by the group velocity

$$v_g = \frac{d\omega}{dk}.$$

Analogous to the (phase) refractive index $n = c/v_{ph}$, a *group* refractive index can be defined according to

$$n' = \frac{c}{v_g} = n + \omega \frac{dn}{d\omega}. \qquad (8\text{–}15)$$

This refractive index can be derived from the cold plasma dispersion relation (Appleton-Hartree formula).

A radio wave propagating in the ionosphere in the presence of a magnetic field is split into two characteristic magnetoionic modes. These modes are reflected (for vertical incidence where $n^2 = 0$) from levels in the ionosphere where the following conditions hold

$$
\begin{aligned}
\text{ordinary mode } (o) \quad & X = 1 && f_o = f_N \\
\text{extraordinary mode } (x) \quad & X = 1 - Y && f_x = \tfrac{1}{2}(f_B + \sqrt{4 f_N^2 + f_B^2}) \\
(z) \quad & X = 1 + Y && f_z = f_x - f_B \, .
\end{aligned}
\tag{8-16}
$$

The total time delay of the reflected signal is given by

$$
\tau = 2 \int\limits_0^R \frac{dR}{v_g}
\tag{8-17}
$$

and thus the observed virtual range P' is related to the true range R by

$$
P' = c \int\limits_0^R \frac{dR}{v_g} = \int\limits_0^R n' \, dR \, .
\tag{8-18}
$$

Since we are interested in the true altitude distribution of electron density, we can rewrite $R = \Delta h = h_s - h$ and $P' = h_s - h'$, where h_s is the height of the satellite (for topside sounding), while for groundbased sounding $R = h$ and $P' = h'$.

In general one can write for the *virtual height* of reflection [180]

$$
h'(f) = \int\limits_{\varphi(f_{N_s})}^{\varphi(f_N)} n'(f, f_N, f_B, \theta) \left(\frac{d \Delta h}{d \varphi} \right) d\varphi = \int\limits_{\varphi(N_s)}^{\varphi(N)} n'(f, N, B, \theta) \, d\Delta h
\tag{8-19}
$$

where $\varphi(N)$ is some single-valued function of the electron density N. (Note: $N(\text{cm}^{-3}) = 1.24 \times 10^4 \, f_N^2 \,(\text{MHz})$ and $f_B (\text{MHz}) = 2.799 \, B \,(\text{gauss})$.)

The derivation of *true height profiles* $N(h)$ from the observed $P' - f$ traces which are called *ionograms* requires for the topside sounder that the plasma frequency and electron gyrofrequency at the satellite be known. Since the topside sounder satellite is immersed in the ionospheric plasma, this information can be obtained from the cutoffs and resonances (cf. Chapter VII) apparent in the topside ionogram as shown in Fig. 59.

The true height analysis can be performed for the ordinary (o) or extraordinary (x) trace using the appropriate refractive index. In practice, the x trace is more frequently used in the analysis of topside ionograms,

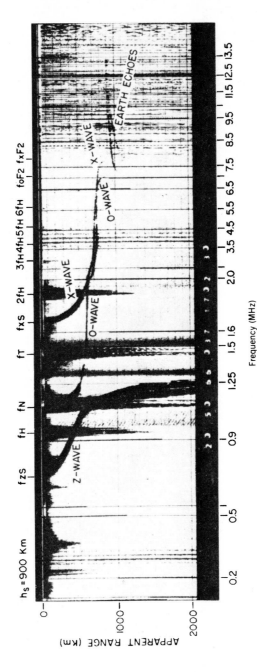

Fig. 59. Topside ionogram obtained with the Alouette I satellite, showing traces for the o, x, and z propagation modes, earth reflections, and cutoffs and resonances at the satellite altitude. (Courtesy of J. F. Jackson)

since this trace always extends to the satellite, whereas the o-trace often does not. The z mode can also be reached from the topside without the need for mode coupling (unlike bottomside observations) because the sounder is within the plasma.

The vertical asymptotes in the ionogram⋆ are called *critical frequencies* since they represent the highest frequency which can be reflected (at vertical incidence) from a given ionospheric layer; frequencies greater than the critical penetrate the layer. The critical frequencies define the maximum density of a given layer; thus $f_o F_2$ and $f_x F_2$ refer to the ordinary and extraordinary critical frequency of the F_2 layer and are related to $N_m F_2$ by virtue of relations (8–16). The large virtual range P' which is characteristic for the critical frequencies is the result of strong signal retardation; i.e., the group refractive index n' is large (the group velocity v_g is small) where the sounding frequency is near the plasma frequency $(X = 1)⋆⋆$ over an appreciable path length as is the case in the vicinity of the layer peaks, or generally where $dN/dz = 0$.

The latter condition may hold even though there is no absolute maximum (peak) in the N_e distribution, but only a *ledge* of relatively constant density. The appearance of large retardation, indicative of a critical frequency in bottomside ionograms (Fig. 60) has led to the

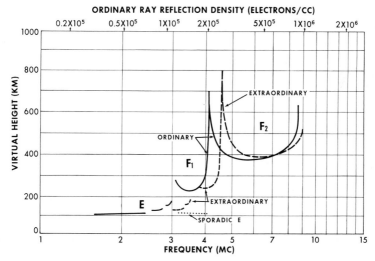

Fig. 60. Bottomside ionogram tracing, showing the critical frequencies for the E, F_1 and F_2 layer (o and x mode), as well as indication of reflections from sporadic E. (Courtesy of J. E. Jackson)

⋆ Those at the satellite altitude (zero range) are called *cutoffs*: $f_o S$, $f_x S$, $f_z S$.
⋆⋆ For the simple case of $n^2 = 1 - X$ it can be shown that $nn' = 1$, i.e., $n' \to \infty$ as $n \to 0$ (cf. 8–18).

concept of an F_1 layer, although in fact this layer is but a ledge in the overall F region electron density distribution.

The true height distribution of N_e is obtained by an inversion of the integral equation (8–19) relating virtual and true heights, which as in the case of the occultation technique is based on an Abel integral transform. The inversion procedure is now exclusively performed by electronic computers using numerical integration methods [180, 181]. In the actual analysis certain assumptions are made for the function $\varphi(N)$, such as linear behavior in N or $\log N$ or polynomial in N or $\log N$, i.e., the inversion is performed by a lamination technique [180]. Fig. 61

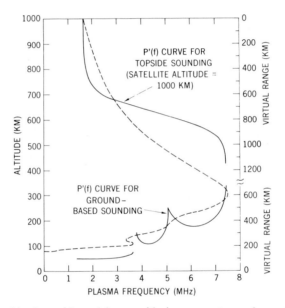

Fig. 61. Combined topside and bottomside ionogram traces (o mode) obtained almost concurrently by a bottomside ionosonde and the Alouette topside sounder satellite and the resulting true height profile of electron density below and above the F_2 peak. (Courtesy of J. E. Jackson)

shows the x traces of a bottomside and a topside ionogram, taken in close proximity to each other, and the resulting electron density profile (in terms of plasma frequency) upon inversion of the $P' - f$ relationship.

Topside sounder satellites (Alouette and ISIS) also provide *in situ* sampling of the ionosphere by exciting plasma resonances which are seen in the ionogram at zero range [182]. They will be discussed briefly in Sect. 2 of this Chapter.

Incoherent Scatter

Ionospheric sounders operate on the principle of total reflection of radio waves which occurs at frequencies of the order of the local plasma frequency $(f \approx f_N \pm \frac{1}{2} f_B)$. However, frequencies $f \gg f_N$ can be scattered by irregularities in the medium (represented by $\Delta N/N$) such as *spread F* and some types of *sporadic E*. In 1958, W. E. Gordon [183] proposed that microscopic irregularities of the ionospheric plasma due to its random thermal motion would scatter radio energy at high frequencies $(f \gg f_N)$ and that high power radars could therefore be used to study the ionosphere even at altitudes not accessible to the conventional ionosondes. Since this scatter is caused by the random thermal motions of the plasma, it is *incoherent* in nature; the backscattered power is proportional to the electron density and the classical Thomson cross section for electrons σ_e, according to

$$P_s \propto \sigma_e N \qquad (8\text{--}20)$$

where $\sigma_e = 4\pi(r_e \sin \psi)^2 = 10^{-24} \, \text{cm}^2$ is the Thomson cross section with r_e the electron radius and ψ the polarization angle, i.e., the angle between incident electric field and observer (for backscatter $\psi = \pi/2$). For this reason, this technique is also sometimes called *Thomson scatter*. The random motions of the electrons lead to a spectrum of the scattered signal which is controlled by the Doppler shift

$$\Delta f_e = \frac{1}{\lambda} \left(\frac{8 \mathscr{k} T_e}{m_e} \right)^{\frac{1}{2}}. \qquad (8\text{--}21)$$

When the feasibility of the incoherent scatter technique was established by K. L. Bowles in 1958 [184], he also found that the spectrum was much narrower than anticipated due to the electron random motions and must therefore be controlled by the ions. Since the ionosphere behaves as a plasma for dimensions greater than the Debye length, only at wavelengths $\lambda < \lambda_D$ is the scattering from free electrons. For wavelengths $\lambda > \lambda_D$, the scattering is from *plasma waves* (ion acoustic waves, electron plasma waves and ion cyclotron waves). The backscatter spectrum, i.e., the displacement from the transmitted frequency, consists of an electron and an ion component and is dominated by the latter as long as the parameter $\alpha = 4\pi \lambda_D/\lambda$, where λ is the wavelength of the incoherent scatter radar, is small*. Only for $\alpha \geq 10$, is the spectrum dominated by the electrons as originally postulated.

---

* The power per bandwidth is much larger for ions than for electrons.

The ion Doppler shift $(\Delta f_i \ll \Delta f_e)$ is given by

$$\Delta f_i = \frac{1}{\lambda} \left(\frac{8 \mathcal{k} T_i}{m_i} \right)^{\frac{1}{2}}. \tag{8-22}$$

As the result of the ion contribution the backscatter cross section is given by [185]

$$\sigma = \frac{\sigma_e}{(1+\alpha^2)\left(1 + \dfrac{T_e}{T_i} + \alpha^2\right)} \tag{8-23}$$

and the backscattered power depends not only on the Thomson cross section but also on the ratio T_e/T_i and α.

The backscatter power can be used to determine the electron density profile if T_e/T_i is known, or in combination with Faraday rotation measurements of the scattered signal which yields N, the ratio T_e/T_i can be deduced.

For small values of $\alpha(<0.3)$, the electron component of the spectrum appears in the *plasma line*, at a Doppler shift from the transmitter frequency according to

$$f_r = \pm f_N (1 + 3\alpha^2)^{\frac{1}{2}}. \tag{8-24}$$

The plasma line is due to scattering from longitudinal electrostatic plasma waves. The plasma line frequency f_r can be interpreted in terms of the Bohm-Gross warm plasma correction to the plasma frequency (see 7–23). Thus, we can express the incoherent scatter parameter α, by

$$\alpha \equiv \left(\frac{k}{\omega_N} v_l \right) = \frac{v_l}{v_{ph}}.$$

For (warm) plasma waves

$$v_{ph} = f_r \lambda_p$$

where λ_p is the wavelength of the plasma wave; furthermore, from (7–6) we have, noting that $2\pi f_r \approx \omega_N$,

$$\alpha = \frac{2\pi v_l}{\omega_N \lambda_p} = \frac{2\pi \lambda_D}{\lambda_p}$$

and since $\alpha \equiv 4\pi \lambda_D / \lambda = 2\pi \lambda_D / \lambda_p$ it follows that $\lambda_p = \lambda/2$.

Thus, the phase velocity of the plasma waves responsible for the plasma line is $v_{ph} = f_r \lambda/2$, where λ is the *radar* frequency. These plasma waves are excited through the Čerenkov mechanism by photoelectrons whose velocity $v \gtrsim v_{ph}$, and whose energy is $E_{ph} = \frac{1}{2} m_e v_{ph}^2$. Landau damping is small as long as (see 7–26), $k\lambda_D < 0.2$, i. e., for $4\pi \lambda_D / \lambda \equiv \alpha < 0.2$.

The plasma line power is enhanced over the thermal level which is proportional to $E_{ph}/\ell\, T_e$. (For negligible Landau damping this ratio has to be $\gtrsim 25$.) Information on the flux and spectrum of photoelectrons can therefore be deduced from plasma line measurements [185].

The spectrum of the backscattered signal for small α is largely determined by the ions (except for the plasma line). On account of the ion Doppler shift (8–22), the *width* of the spectrum varies as $(m_i)^{-\frac{1}{2}}$. The normalized spectra for different ions have approximately the same shape and it is not possible to determine either T_i or m_i if the other is unknown (Fig. 62). In an ion mixture of comparable abundance, the spectrum is

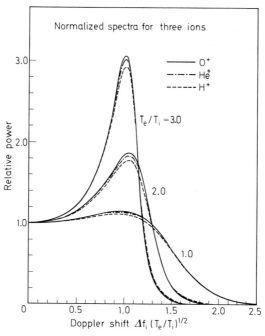

Fig. 62. Incoherent backscatter radar spectrum for O^+, He^+ and H^+ and different ratios of T_e/T_i; the Doppler shift Δf_i corresponds to that of an ion travelling toward the radar at mean thermal speed. (After Evans [186])

not simply the sum of the spectra of the individual ions. However, if T_e/T_i and α are known, a unique interpretation of the measured spectra in terms of ion composition is possible.

For propagation perpendicular to the magnetic field, ion-cyclotron waves are also excited and ion cyclotron resonances (Ω_B) occur when the radar wavelength is of the order of the ion cyclotron radius at displacements from the transmitter frequency corresponding to

$$\Delta f = \frac{1}{2\pi} n \Omega_B \left(2\pi + \frac{\Omega_B}{\sqrt{\pi \Delta f_i}} \right). \qquad (8\text{--}25)$$

Plasma drifts can also be determined from the incoherent scatter spectra as the result of a shift in the spectrum according to

$$\Delta f_d = \frac{+2v_d}{\lambda} \qquad (8\text{--}26)$$

where v_d is the drift velocity and the $+$ sign refers to the direction towards the radar [186].

In the region where ions and neutrals are collisionally coupled, the ion properties can also be used to infer characteristics of the neutral atmosphere. Thus, the incoherent scatter radar has become not only the most "powerful" (transmitter power $\sim$ Megawatts) but also the most versatile technique for the investigation of the terrestrial ionosphere. However, on account of its physical size, the incoherent scatter radar must unfortunately remain an earthbound technique.

Although the high-power incoherent scatter radar does not perceptibly perturb the ionosphere by virtue of its high frequency ($f \gg f_N$), important modifications of the ionospheric medium have been observed with high-power radio waves at frequencies $f \gtrsim f_N$ [187].

VIII.2. Direct Measurement Techniques

Rockets and satellites have made possible the direct *(in situ)* measurement of ionospheric parameters by means of plasma probes. The *in situ* plasma measurements are based either on the *Langmuir* probe method, which has been used for many years in the laboratory to determine plasma density and temperature, or on *plasma resonance* phenomena; the former are essentially *dc* methods whereas the latter are *rf* methods.

In-situ dc Techniques

Langmuir probe techniques utilize probes of different geometrical configurations, from simple cylindrical (wire) probes to planar and spherical *traps* (Faraday cups) having in addition to the collector a number of grids which are appropriately biased to exclude unwanted current components. However, all of them are based on the application of Langmuir probe theory which is valid as long as the probe dimensions are large compared to the Debye length [188].

From the *I–V* (current-voltage) characteristic which is obtained by varying the probe potential V_p, the density and energy distribution (temperature) of the charged particles can be determined. A typical *I–V* curve obtained with a Langmuir probe onboard of a satellite is shown in Fig. 63. The potential of the probe with respect to the plasma is $V = V_p - V_0$, where V_0 is the plasma potential. For a Maxwellian distribution the measured electron current as a function of V can be expressed, for the *retarding region* ($V < 0$, i. e., a potential of the same sign as the particle charge) by*

$$I_e = I_0 \exp\left(\frac{eV}{\mathscr{k}T_e}\right) \qquad (8\text{–}27)$$

where $I_0 = e N A (\mathscr{k}T_e / 2\pi m_e)^{\frac{1}{2}} \equiv N e A \bar{v}_e / 4$ where N is the electron density, T_e is the electron temperature and A is the probe area; I_0 also

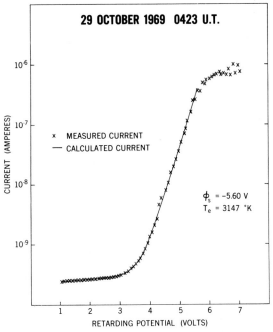

Fig. 63. $I - V$ curve obtained by a satellite-borne Langmuir probe; the slope is indicative of the electron temperature and the electron density is related to the current at plasma potential $(-\phi_s)$ where the $I - V$ curve starts to depart from an exponential

---

* In the presence of a magnetic field, however, the electron collection by a probe may be reduced [188a].

corresponds to the random electron current, with $\bar{v}_e = (8\pounds T_e/\pi m_e)^{\frac{1}{2}}$ the *average* velocity of the electrons. The current density $j_e = I_e/A$ represents the *integral flux* for energies $E > eV$.

Whereas the current in the retarding range is independent of probe *shape*, this is not the case for accelerating potentials. The value I_0 is the current at the plasma potential at which *both* electrons and ions are collected without the aid of an accelerating potential. This potential can be identified by the break in the I–V curve. The fact that $V_p = V_0$ does not occur at $V_p = 0$ results from the fact that a spacecraft (probe) immersed in the ionospheric plasma acquires a potential which has to be compensated for. The spacecraft potential is given by $\phi_s \equiv V_s = -V_0$, i.e., it can be identified by the breakpoint between retarding and accelerating regions. The equilibrium potential (sometimes called floating potential) for a body at rest is given by the condition that the net current is zero and thus

$$V_s = -\frac{\pounds T_e}{e} \ln\left(\frac{I_{oe}}{I_{oi}}\right), \tag{8–28}$$

i.e., it results from the random currents of the electrons and ions and is therefore dominated by the electrons. However, in reality the spacecraft potential is modified by photoemission, energetic particle fluxes as well as a $v_s \times B$ contribution especially when the spacecraft has long booms [189].

The electron density can be determined from I_0, according to

$$N = \frac{I_o}{Ae}\left(\frac{2\pi m_e}{\pounds T_e}\right)^{\frac{1}{2}} \tag{8–29}$$

and the electron temperature from the logarithmic slope of the I–V characteristic according to

$$T_e = -\frac{e}{\pounds}\left[\frac{d(\ln I_e)}{dV}\right]^{-1} \tag{8–30}$$

Departures from a Maxwellian distribution are exhibited by a *non-linear* relationship between $\ln I_e$ and V in the retarding region.

Actual electron energy distributions can be derived by means of the *Druyvesteyn* analysis [188]

$$f(E) = \frac{1}{N}\frac{dN}{dE} = \frac{(8m_e)^{\frac{1}{2}}}{Ae^{\frac{3}{2}}}V^{\frac{1}{2}}\frac{d^2 I}{dV^2}. \tag{8–31}$$

Whereas the electron density at plasma potential depends on T_e, a simple cyclindrical Langmuir probe determines the electron density in the acceleration region without *a priori* knowledge of T_e. Because of the cylindrical geometry (actually a thin wire), the electron current in the accelerating region (for $eV/\not{k}T_e \geq 5$) is given by

$$I_e = NeA\left(\frac{2eV}{\pi^2 m_e}\right)^{\frac{1}{2}}$$

and T_e can again be determined in the retarding region.

Probes of planar or spherical geometry, usually employing grids with appropriate potentials to remove unwanted currents are frequently used; they are called *retarding potential analyzers (RPA)* and are also used to measure suprathermal particle fluxes.

Langmuir theory also applies to *ion* measurements [188, 189]. The principal difference is that for ions the motion of the spacecraft through the ionosphere can usually not be neglected; i.e., the spacecraft is usually supersonic relative to the ions. The random ion current is also reduced by $(m_e/m_i)^{\frac{1}{2}}$ and ion measurements may be masked by photo-emission currents. For an *ion trap (RPA)* the current is enhanced above the Maxwellian random current as a function of the ratio of spacecraft velocity to the most probable velocity of the ions, v_s/c_i. For a ratio $v_s/c_i < 0.25$, the ion current is close to the random current, while for a ratio $v_s/c_i \geq 2.5$ the current is nearly independent of the thermal speed and the controlling factor is the spacecraft velocity.

For planar geometry, the general equation that applies for positive ion current is given according to [cf. 189]

$$I_+ = \alpha e N_i A v_s \cos\psi\left[\left(\frac{1}{2} + \frac{1}{2}\operatorname{erf}(X) + \frac{c_i\exp(-X^2)}{2\sqrt{\pi}}\right)\right] \quad (8\text{--}32)$$

where $X = (v_s\cos\psi)/c_i - \sqrt{eV/\not{k}T}$, α is the transparency of the grids and ψ is the angle between the spacecraft velocity vector and the trap normal. (A similar, though slightly more complex formula applies to spherical geometry [cf. 188].)

For the special case where $\psi < 45°$ and the spacecraft is supersonic with respect to the ions, the ion current simplifies to

$$I_+ = \alpha A e N_i v_s \cos\psi. \quad (8\text{--}33)$$

Thus, I_+ is a maximum in the ram direction ($\psi = 0$) and approaches zero when the trap normal is perpendicular to this direction. An ion retarding potential analyzer provides information on *ion energy*, and can therefore yield information on ion composition and temperature, since the I–V curve depends on T_i and the ion masses and the relative abundances of the ionic constituents. These parameters can be obtained by multi-parameter curve fitting, assuming *a priori* knowledge of the types of ions from equation (8–32). A typical I–V curve for more than one ionic constituent is shown in Fig. 64.

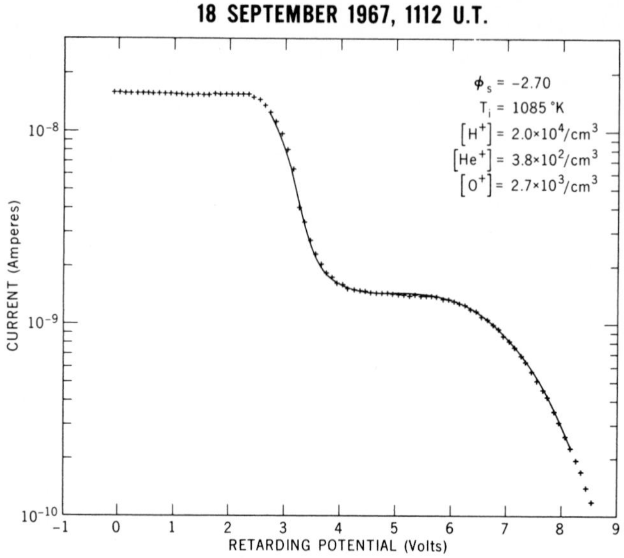

18 SEPTEMBER 1967, 1112 U.T.

Fig. 64. $I-V$ curve obtained by an ion retarding potential analyzer (RPA) on OGO 4, showing the presence of the ions H^+, O^+ and He^+

More directly, the ion composition (ion mass) and ion temperature can be determined by a Druyvesteyn analysis, according to

$$f(E) \propto \frac{m_i^{\frac{1}{2}} V^{\frac{1}{2}} d^2 I_+}{dV^2}. \tag{8-34}$$

For a multi-constituent ionosphere the distribution function exhibits nearly Gaussian peaks; the position of the center of the peaks is related to the mean energy of ions of a given mass and occurs where

$$E = eV = \tfrac{1}{2} m_i v_s^2.$$

The width of the peaks at the $1/e$ point is

$$\mathscr{W} = 4(\mathscr{k}\, T_i \cdot \tfrac{1}{2} m_i v_s^2)^{\frac{1}{2}}$$

where v_s is the spacecraft velocity corresponding to the case $v_s/c_i \gg 1$.

Hence, both the position and width of the peaks provide information on the *ion mass*, while the width also gives information on the ion temperature (Fig. 65). The ion concentration is given by the first derivative, according to $dI/dV \propto \int f(E)dE \equiv N$.

While RPA's give information on ion temperature and composition, they represent only relatively low-resolution devices for ion composition

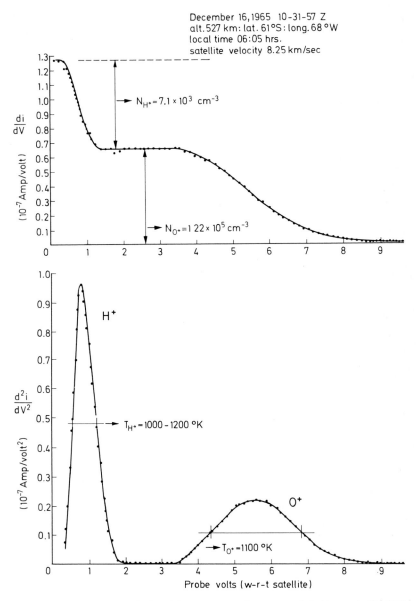

Fig. 65. Druyvesteyn analysis of ion currents, in terms of dI/dV and d^2I/dV^2. The derivatives are obtained electronically onboard the satellite by means of an a.c. technique. (After [218])

measurements. For high resolution of ion masses, *ion mass spectrometers* are employed [190]. Ion mass spectrometers generally used on spacecraft are of the following types [191]:

1) The *rf (Bennett) mass spectrometer* belongs to the class of time-of-flight spectrometers, i. e., ions of different charge-to-mass ratio are separated by virtue of the time they require to travel a given path length under the influence of (retarding and accelerating) dc and rf potentials.

2) The *magnetic deflection mass spectrometer* employs acceleration of ions by an electrostatic field and subsequent deflection by a magnetic field through an angle of 60 (90, 120) degrees. The ions of different m_i/e will have different radii of curvature for a given accelerating potential; for a given radius of curvature, ions can be distinguished by varying the accelerating potential.

3) The *quadrupole mass spectrometer* is based on the concept that a hyperbolic field configuration due to dc and rf voltages applied to four equally spaced (circular or hyperbolic) rods allows only ions of one mass to have a stable path near the axis and be collected by the ion collector. Ions whose masses are not compatible with the given voltages follow trajectories which make them strike the rods and thus prevent then from reaching the collector. This device represents a mass filter and is also known as *Massenfilter*.

A problem common to all ion mass spectrometers is the conversion of measured collector currents to ion densities. This conversion is commonly performed by resorting to laboratory calibrations or by normalizing to total densities obtained by other devices. Generally, the current at the collector is given by

$$I_c = \frac{\kappa N_i e c_i A}{2\sqrt{\pi}} f\left(E_a, \frac{v_s}{c_i}\right) \qquad (8\text{--}35)$$

where κ is the spectrometer efficiency, A is the aperture area, c_i is the most probable thermal velocity of the ion of mass m_i and $f(E_a, v_s/c_i)$ is a function of the attracting potential (used to bring ions into the aperture and keep out electrons) and the ratio of spacecraft-to-ion velocity. The collector current is greatly enhanced when $E_a \gg \ell T_i$ and $f(E_a, v_s/c_i)$ is then much greater than unity; this function is often lumped into an effective area $A_{eff} \equiv A \cdot f(E_a, v_s/c_i)$.

In-situ rf Techniques

In contrast to Langmuir probes which draw a *dc* current from the ionospheric plasma and whose *I–V* characteristics lead to information on plasma density and temperature, *in situ rf-techniques* determine the characteristics of the ionosphere from its dispersive properties. The latter usually refer to spatial averages over a much larger volume than is the case for Langmuir probe techniques.

The *rf impedance probe* is based on the fact that the impedance of an electrically short antenna at frequencies $f \gg f_N$, f_B is a capacitive reactance $(1/\omega C_0)$ and departs from the free space value C_0 by an amount ΔC due to the presence of the ionospheric plasma $(n = 1 - X)$, according to [192]

$$\frac{\Delta C}{C_0} = K \frac{N_e}{f^2} \qquad (8\text{--}36)$$

where $K = 80.6$ for N in cm^{-3} and f in kHz.

For the general case, including the effect of a magnetic field and the ion sheath formed around the probe, the probe capacitance within the magneto-active plasma is given by

$$C = C_0 \left(1 + \frac{X(1 + \mathscr{G}(\psi) Y^2)}{1 - Y^2}\right) \qquad (8\text{--}37)$$

where $\mathscr{G}(\psi)$ is a factor which depends on the probe geometry and the angle between the electric field of the rf wave and the magnetic field, and X and Y are the magnetoionic parameters. By varying the frequency applied to the rf probe, two series-resonances maxima and two parallel-resonance minima are obtained. From these resonances all unknowns, including ω_B, ω_T and ω_N, and hence the electron density can be determined. A variety of these rf probes exists, operating in different magneto-ionic regimes [174].

As long as $0 < X < 1 - Y$, the impedance is represented by a reactance greater than its free space value. When the upper hybrid frequency $(\omega_T; X = 1 - Y^2)$ is crossed, the resistive component of the impedance can become large and the reactance changes from capacitive to inductive. Between f_N and f_T large increases in the real (resistive) part of the antenna impedance occur, which seem to be associated with enhancements in radio noise observed with sensitive receivers on satellites [193]. This noise enhancement may be the result of plasma wave generation by the Čerenkov process; the cutoffs of the noise bands again can be used to determine local electron density [194].

Similarly, the low-frequency cutoff of VLF noise observed on satellites occurs at the *lower hybrid resonance* frequency

$$f_{LHR} = \left[\frac{f_N f_B}{f_T}\right] \left[\frac{m_e}{m_p} + \frac{1}{\tilde{m}_i}\right]^{\frac{1}{2}} \qquad (8\text{--}38)$$

where $\tilde{m}_i$ is the harmonic mean of the mass numbers of all positive ions $1/\tilde{m}_i = \sum \eta_i (m_p/m_i)$, with $\eta_i = n_i/N$ the relative abundance of the particular ions (whereas the mean ion mass is given by $m_i = \sum \eta_i n_i$) and m_e/m_p is the ratio of electron to proton mass. Thus, if the other plasma parameters f_N, f_T, f_B are known, the relative abundance of ionic species can be determined [195].

Ion cyclotron whistlers observed with VLF receivers on satellites can also be used to determine ion composition from the "cross-over" frequency between electron and ion whistler, and the ion temperature from the difference between the ion whistler at frequency cutoff and the asymptotic ion cyclotron resonance frequency [195].

The plasma resonances observed on the topside sounder satellites represent an extremely useful tool for determining the ionospheric plasma parameters [182]. These resonances are excited by the rf field of the sounder and represent electrostatic plasma waves whose wave normals are either nearly parallel to the magnetic field $(\boldsymbol{k} \times \boldsymbol{B} \approx 0)$, as is the case for the f_N and f_B resonances, or nearly perpendicular to the magnetic field $(\boldsymbol{k} \cdot \boldsymbol{B} \approx 0)$, as is the case for the upper hybrid resonance f_T and the electron-cyclotron harmonics $n f_B$ (Fig. 66). Since the plasma resonances correspond to plasma waves in the "warm" ionspheric plasma, these waves can also propagate [163]. Quasiperiodic amplitude fluctuations (fringes) in the plasma resonance f_N and the upper hybrid resonance f_T have been explained as "beating" of two waves having dif-

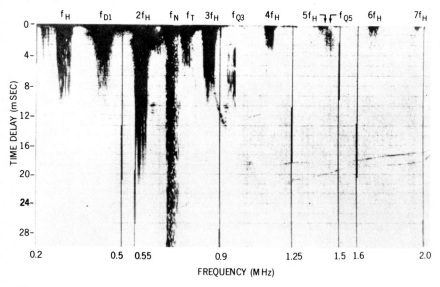

Fig. 66. Ionospheric plasma resonances and cutoffs appearing (at zero range) in an ionogram obtained with the Alouette II satellite. (Courtesy of R. F. Benson)

ferent group paths due to small plasma density gradients (non-uniformity of the ionospheric plasma). This beat frequency corresponds to

$$f_b \simeq \frac{v_s}{4\pi} \left(\frac{\omega' \omega_N t}{H v_l} \right)^{\frac{1}{2}} \tag{8-39}$$

where v_s is the satellite velocity, t is the observed time delay of the beat, H is the scale height for the electron density and $\omega' = \omega_N + \varepsilon t$ is a frequency close to ω_N (ε small) and v_l is the (longitudinal) thermal velocity of the electrons. Hence,

$$f_b \propto N^{\frac{1}{2}} T_e^{-\frac{1}{4}} \tag{8-40}$$

and it is therefore possible to determine the average electron temperature along the group path from this phenomenon [196].

Another possibility for determining electron temperature was suggested to be from the two-component nature of some of the diffuse resonances f_{D_n} [197]. This feature seems to arise from the Doppler-shift of electrostatic cyclotron waves propagating perpendicular to the magnetic field (where Landau damping is absent). From the propagation characteristics of these waves it is again possible to infer the electron temperature.

Currently, a discrepancy seems to exist between plasma temperatures (especially the electron temperature) determined by Langmuir probes and those obtained by incoherent backscatter radar which is based on plasma wave phenomena. Since the probe and backscatter temperatures are determined from different moments of the electron velocity distribution function $f(v)$, a departure from a Maxwellian or an isotropic distribution could lead to different values for the electron temperature [198]. The thermodynamic electron temperature, for which the probe temperatures are representative, is based on the *average* energy of the entire electron population according to

$$\tfrac{3}{2} \cdot \mathscr{k} T_{th} = \int \tfrac{1}{2} m v^2 f(v) dv \tag{8-41}$$

whereas the shape of the incoherent backscatter spectrum (based on the ion component), depends on the electron energy distribution by virtue of

$$\frac{m}{\mathscr{k} T_b} = \int \frac{1}{v^2} f(v) dv. \tag{8-42}$$

The same is true for other plasma wave propagation phenomena, since the warm plasma correction also involves the (longitudinal) electron thermal velocity $v_l^2 = \mathscr{k} T / m$. If the electron distribution function is distorted at low energies, more weight is given to these low energy electrons which could lead to the appearance of an electron temperature lower than that based on the average energy.

Chapter IX

Observed Properties of Planetary Ionospheres

IX.1. The Terrestrial Ionosphere

Information on the ionosphere of Earth has accumulated for the past four decades. Originally, electron density was the only primary quantity observable by groundbased radio techniques. During the last two decades information on other physical parameters of the ionosphere, such as charged particle temperature (T_e, T_i) and ion composition (m_i), as well as measurements of the ionizing radiations and the neutral atmosphere have become available. Satellites have provided a *global* picture of these parameters while at selected locations, groundbased incoherent back-scatter radars have made possible the study of the *time* variations of these parameters as a function of altitude. In addition, sounding rocket experiments have provided direct measurements of many ionospheric parameters as a function of altitude, especially in the lower ionosphere.

The Lower Ionosphere

The term is used for the *D*- and *E-region*, both of which are primarily controlled by photochemical rather than by plasma transport processes.

   The formation of the *D-region* is the result of the following ionization sources: i) Solar Lyman α (1216 Å) ionizing the minor constituent NO, ii) solar X-rays ($\lambda < 8$ Å) ionizing N_2, O_2 and A, iii) galactic cosmic rays ionizing all atmospheric constituents and iv) photoionization of the metastable O_2 ($^1\Delta_g$) by solar UV radiation ($\lambda < 1118$ Å) [199, 200]. The relative importance of these ionization sources for the quiet day-time D-region is illustrated in Fig. 67, while electron density profiles observed in the D-region by sounding rockets are shown in Fig. 68. These profiles also illustrate the effect of the variation of the X-ray flux in the range $\lambda < 8$ Å. The lowest density profile (NASA 14.107) corresponds to a completely quiet sun with an X-ray flux (1–8 Å) of 1.9×10^{-4} erg cm^{-2} sec^{-1}, while the highest density profile corresponds to a Class 1 flare with an X-ray flux of 9×10^{-2} erg cm^{-2} sec^{-1} with the intermediate profiles representing a somewhat enhanced X-ray flux [200].

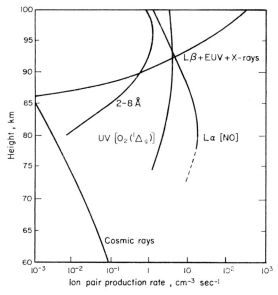

Fig. 67. Ion production rates for the main ionization sources of the lower iono-sphere. (After Thomas [199])

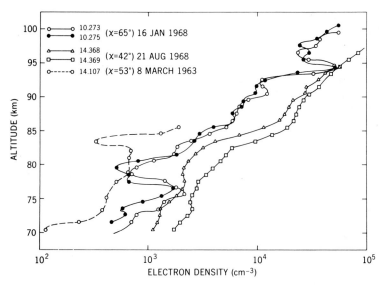

Fig. 68. Electron density profiles in the lower ionosphere obtained by rocket measurements illustrating the effect of a variation in the X-ray flux ($\lambda < 8$ Å) on the D-region ionization. Profile NASA 14.107 corresponds to a completely quiet sun while NASA 14.369 corresponds to Class I solar flare. (After A. C. Aikin [200])

During the night, the electron density below 85 km is $\sim 10^2$ cm^{-3}, mainly due to cosmic ray ionization; a distinct "C-layer" is sometimes seen at an altitude of 65—68 km. When the solar zenith angle reaches $\chi = 95°$, the ionization starts to increase, following the expected solar zenith angle control for a Chapman layer, i. e., $N \propto \cos^{\frac{1}{2}} \chi$. In addition to cosmic ray ionization, electron precipitation ($E > 40$ keV) is thought to be a possible source of the nighttime D-region at low and middle latitudes. The so-called winter anomaly in ionospheric absorption which is due to enhanced D-region ionization is considered to be the combined result of seasonal variations in NO and particle precipitation. Increases in D-region ionization during magnetic storms appear to be associated with enhancement in NO, which in turn arises from meteorological phenomena in the mesosphere that are suspected to be due to magnetic storm effects. As indicated before, the D-region ionization is greatly enhanced during solar flares as the result of the increased X-ray flux between 1 and 8 Å. Short duration (10 -20 min) flare effects are known as *sudden ionospheric disturbance (SID)* or *Moegel-Dellinger effect.* High fluxes of solar protons in the energy range from 10 to 100 MeV produced in the course of a large solar flare cause greatly enhanced ionization in the lower D-region where the collision frequency is high. This enhanced ionization leads to strong radio wave absorption over the polar cap to (dip) latitudes of 65°, which have called *polar cap absorption (PCA) events* [cf. 174]. Associated with the magnetic storm following the flare, the polar cap absorption region extends to even lower latitudes ($\sim 55°$). Strong enhancement of D-region ionization also occurs as the result of ionizing radiations associated with nuclear explosions.

For many years the ion composition of the D-region has been the subject of speculation. Mass-spectrometric measurements at these low altitudes are difficult due to the relatively high pressure. Recently, however, rocket-borne pumped ion mass spectrometers have been successful in identifying ions in the D-region. The results were most surprising, since they showed that instead of the expected NO$^+$ ion resulting from Ly α ionization of NO, and O$_2^+$ resulting from ionization of O$_2$ and N$_2$ (N$_2^+$ being rapidly converted into O$_2^+$) by X-rays and galactic cosmic rays, heavy hydrated water cluster ions predominate in this region [201]. The originally produced ions (NO$^+$ and O$_2^+$) convert to water clusters of the type H$_3$O$^+ \cdot$ (H$_2$O)$_n$ mainly by three-body reactions. Examples of recent rocket measurements with a pumped quadrupole ion mass spectrometer [202] are shown in Fig. 69. There seems to be evidence, that the heavier water cluster ions are dissociated by the shockwave produced by the rocket. A reaction sequence proposed to explain the observations in Fig. 69 is shown in Fig. 70.

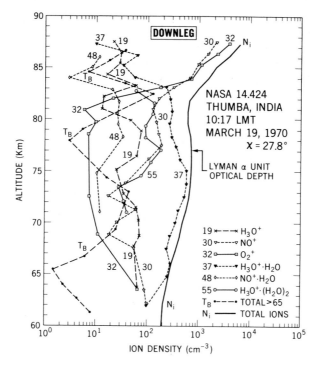

Fig. 69. Ion composition in the equatorial lower ionosphere measured by a rocket-borne pumped quadrupole ion mass spectrometer, illustrating the predominance of heavy hydrated water cluster ions in the D-region. (After [202])

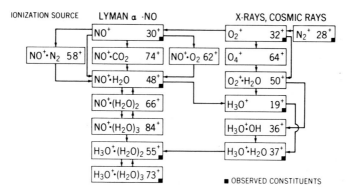

Fig. 70. Reaction scheme leading to cluster ions observed in the D-region as illustrated in Fig. 69. (After Goldberg and Aikin [202])

An even more difficult problem than that of the positive ion composition is the composition of negative ions [203, 204]. In the D-region, negative ions are present as the result of various, but primarily three-body attachment processes. The initial reaction is the attachment of an electron to O_2 according to

$$e + O_2 + O_2 \rightarrow O_2^- + O_2 \qquad (9\text{-}1)$$

followed by other steps illustrated in Fig. 71. In addition to the reactions shown in Fig. 71, the negative ions can become hydrated as easily as the positive ions leading to negative ion water clusters. Negative ions have now also been measured by a pumped ion mass spectrometer [205].

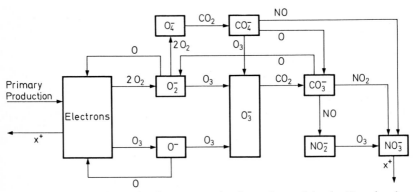

Fig. 71. Reaction scheme leading to negative ions observed in the D-region by rocket-borne ion mass spectrometer. (After [205])

At present, one of the most challenging problems in the understanding of the terrestrial ionosphere is the ion chemistry of the D-region and its relation to the minor constituents of the atmosphere such as H, NO, NO_2, H_2O, O_3, CO_2 and the possible links between the mesosphere and stratosphere.

It is generally accepted that the ionospheric *E-region* is produced by X-rays in the range from 10 to 100 Å and by EUV radiation in the range from 800 Å to Lyman-beta 1026 Å. The relative importance of the X-rays, however, has been a matter of some controversy. It appears now that, at least near solar minimum, the X-ray intensity is probably too small to contribute to the formation of the E-region, so that the main ionization source should then be EUV radiation $\lambda < 1026$ Å forming O_2^+.

Ion composition measurements show that NO^+ and O_2^+ are the major ions in the E-region, with N_2^+ and O^+ as minor ions. The NO^+ ion arises from the fast ion-molecule reaction $O_2^+ + N_2 \rightarrow NO^+ + N$; the

N_2^+ produced by ionizing radiations are efficiently transformed into O_2^+ by charge exchange (cf. [IV.3]).

The E-region conforms closely to an ideal Chapman layer. The base of the E-region is at about 90 km, i. e., where the ion production rate due to Lyman-beta 1026 Å exceeds the ion production of NO^+ by Lyman-alpha. Near the base of the E-region is a metal ion layer (Mg^+, Ca^+, Fe^+, Na^+, Si^+) resulting from neutral metal species of meteoric origin. Rocket measurements of the mid-latitude E-region ion composition [206] are illustrated in Fig. 72.

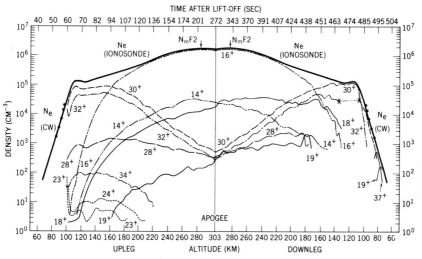

Fig. 72. Ion composition of the E and F region measured by a rocket borne quadrupole ion-mass spectrometer. (After Goldberg and Blumle [206])

The E-region is centered at ~ 105 km, remaining relatively constant, while the maximum electron density of the E layer varies with solar zenith angle according to $N_m E \propto \cos^{\frac{1}{2}} \chi$. However, even during the night, the E layer persists, having a nearly constant density of 10^3 to 10^4 cm^{-3}. This nighttime E layer is now thought to be maintained by scattered EUV radiations, primarily Lyman-beta 1026 Å and Lyman-alpha 1216 Å [199, 200]. Within the E-region, local enhancements and irregularities in electron density, known as *sporadic E* (E_s), are observed [207]. At midlatitudes sporadic E is the result of layering of metal ions due to an electric field produced by wind shear, together with the long lifetime of metallic ions; this type of sporadic E exhibits sharp gradients of electron density with height. The equatorial sporadic E, on the other hand, is due to small-scale irregularities resulting from the two-stream plasma instability occurring in the electrojet region. In the auroral zone, the NO^+ density (and O^+) is found to be greatly enhanced compared to that at midlatitudes; this seems to be associated with an increase of neutral NO and its direct photoionization. The charged particles temperatures in the E-region seem to be close to the neutral gas temperature, except possibly in the auroral zone. Table 32 shows positive and negative ions detected in the *lower* ionosphere.

Table 32. *Ions Detected in the Lower Ionosphere* ★

Positive Ions		Negative Ions	
Identification	Mass Number	Identification	Mass Number
N^+	14		
O^+	16	O_2^-	32
H_3O^+	19	Cl^-	35, 37
Na^+	23		
Mg^+	24, 25, 26	CO_3^-	60
Al^+	27	HCO_3^-	61
Si^+	28	NO_3^-	62
N_2^+	28		
NO^+	30	$O_2^- \cdot (H_2O)_2$	68
S^+	32, 34	CO_4^-	76
O_2^+	32, 34	$CO_3^- \cdot H_2O$	78
$H_3O^+ \cdot H_2O$	37	$NO_2^- \cdot (HNO_2)$	93 ± 1
K^+	39, 41	$CO_4^- \cdot (H_2O)$	
Ca^+	40	$NO_2^- \cdot (HNO_2) \cdot H_2O$	111 ± 1
Sc^+	45		
NO_2^+	46	$CO_4^- \cdot (H_2O)_2$	
$NO^+ \cdot H_2O$	48	$NO_3^- \cdot (HNO_3)$	125 ± 1
Cr^+	52		
Mn^+	55		
$H_3O^+ \cdot (H_2O)_2$	55		
Fe^+	56, 54		
Ni^+	58, 60		
HNO_3^+	63		
$H_3O^+ \cdot (H_2O)_3$	73		
$HNO_3^+ \cdot (H_2O)$	81		
$H_3O^+ \cdot (H_2O)_4$	91		
$H_3O^+ \cdot (H_2O)_5$	109		
$H_3O^+ \cdot (H_2O)_6$	127		

★ Prepared by A. C. Aikin.

The F Region and Topside Ionosphere

The ionospheric F-region is historically subdivided into the F_1 and F_2 layer. The F_1 *ledge* is representative of the *maximum of ion production,* whereas the F_2 *peak* representing the *maximum of electron density* is the result of the combined effects of ion-chemistry and plasma diffusion [208]. The ionization sources for the F-region are solar EUV radiations in the Lyman continuum (910—800 Å), in the range 350—200 Å, including the strong He II (304 Å line, and additional contributions from the range 500 to 700 Å. The maximum absorption occurs in the altitude range 160 to 180 km, with only a small fraction (10%) of these radiations penetrating below 145 km. These ionizing radiations are responsible for the formation of O_2^+, N_2^+, O^+, He^+, N^+, while the actually observed ion composition which results from chemical processes differs from the primary ones. In the F_1 region the major ions are NO^+ and O_2^+ (with O^+, N^+ the minor ions); in the F_2 region O^+ is the principal constituent, with H^+, N^+ and He^+ the minor constituents. A vertical profile of the ion composition and its diurnal behavior is shown in Figs. 73 and 74. The F_1 region is controlled by photochemical equilibrium with a square loss law due to dissociative recombination. This region closely resembles a simple Chapman layer.

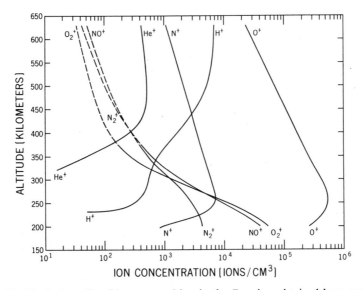

Fig. 73. Vertical profile of ion composition in the *F* region obtained by a rocketborne r. f. ion mass spectrometer. In addition to the major ion O^+, molecular ions O_2^+, NO^+, are seen at lower altitudes and the minor ions N^+, H^+, He^+ at the higher altitudes. [After Brinton et al.; J. Geophys. Res. 74, 2944 (1969)]

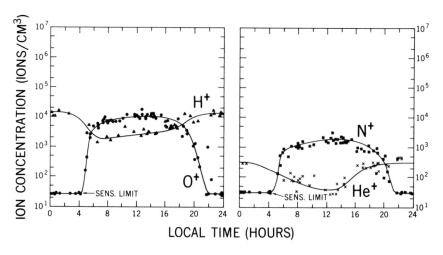

Fig. 74. Diurnal variation of ions in the topside ionosphere obtained with an rf ion mass spectrometer on Explorer 32 showing the anticorrelation of light ions (H^+, He^+) and N^+ and O^+. (After Brinton et al.: J. Geophys. Res. 74, 4064 (1969))

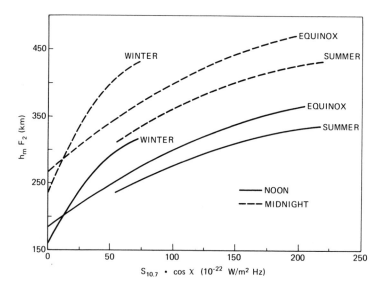

Fig. 75. Variation of the height of the F_2 peak ($h_m F_2$) for a midlatitude location (Lindau/Harz) as function of solar activity expressed in terms of 10.7 cm solar radio flux ($S_{10.7}$) corrected for solar zenith angle. (After W. Becker [211a])

In the F_2 region the controlling chemical loss process is the ion-molecule reaction converting O^+ into molecular ions (NO^+ and O_2^+) which ultimately recombine dissociatively; this loss process obeys a linear loss law, with a loss coefficient β which decreases with altitude. As the result of this process the electron density increases, i. e., the F_2 region resembles a "Bradbury" layer. The F_2 peak is formed by the combined action of this linear loss process and plasma diffusion at an altitude where the chemical time constant $\tau_C = 1/\beta$ is about equal to the plasma diffusion time constant $\tau_D \cong H_D^2/D_a$. In addition to these

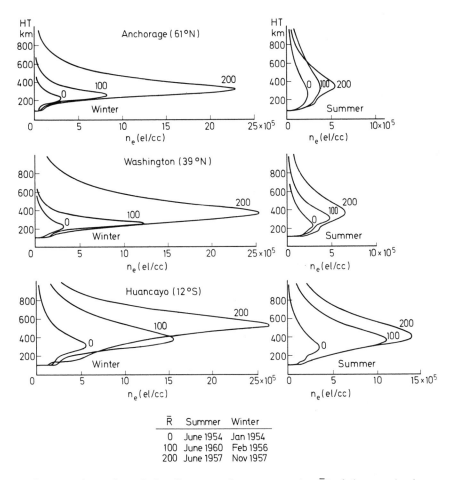

$\bar{R}$	Summer	Winter
0	June 1954	Jan 1954
100	June 1960	Feb 1956
200	June 1957	Nov 1957

Fig. 76. Solar cycle variation (in terms of sunspot number $\bar{R}$) of electron density profiles for selected seasons and locations (Topside is extropolated by means of a Chapman distribution). (After J. W. Wright, [212])

processes, the height of the F_2 peak is also affected by the vertical components of horizontal winds or plasma drifts perpendicular to the magnetic field resulting from electric fields [209, 210, 211].

The height of the F_2 peak, $h_m F_2$, varies between 200 km and 450 km. It is lower during the day than during the night and lower in summer than in winter. The height of the F_2 peak is greater during periods of high solar activity and higher over the equator than over midlatitudes. Fig. 75 shows $h_m F_2$ for a midlatitude station as a function of solar activity (expressed in terms of the 10.7 cm solar radio flux adjusted for the solar zenith angle $\cos \chi$), based on observation from a solar maximum to a solar minimum [211 a].

The maximum density, $N_m F_2$ does not follow a simple $\cos \chi$ dependence due to the fact that not only photochemical but also plasma transport processes determine its value. The day-night variation is about a factor of 10, as is the variation over a solar cycle for the winter values, while for the summer values it is only about a factor of three. This solar cycle variation is illustrated in Fig. 76 [212].

The departures from a simple solar control (i.e., Chapman layer behavior) for the F_2 region are referred to as *anomalies*. The *seasonal anomaly* refers to the fact that the electron density is higher in (local) winter than in summer, which seems to be related to the composition and dynamics of the neutral atmosphere, particularly the ratio O/O_2 and O/N_2 which controls ion production and loss in this region. This "winter anomaly" is actually due to a depression in electron density in summer due to the abundance of O_2 in the summer hemisphere as the result of the global atmospheric circulation. The other seasonal anomaly refers to the fact that the plasma density exhibits a semiannual variation with maxima in March and October and minima in July and January. This "semiannual" variation in N_e seems to be related to such a variation in the O/O_2 ratio which shows maxima in the equinoctial months and a variation by a factor of two from equinox to solstice [213]. Thermospheric winds also play a role in the maintenance of the nighttime F region [210]. These winds raise the height of the F_2 peak causing a decrease in the loss rate of O^+. In the F region, there is also absence of thermal equilibrium, with the electron and ion temperatures exceeding the neutral gas temperature [214, 215, 216, 217]. With the decrease in electron temperature near sunset, ionization stored in the *topside ionosphere (protonosphere)* rapidly diffuses downward helping to maintain the nighttime F region [211]. Typical seasonal and diurnal variations of $N_m F_2$ and $h_m F_2$ at midlatitudes are illustrated in Fig. 77. Diurnal variation of the electron temperatures in the topside ionosphere [216] are shown in Fig. 78. The ratio of T_e/T_i during the night is $T_e/T_i \cong 1$, while during the day, $T_e/T_i > 1$ [218]. The global variation

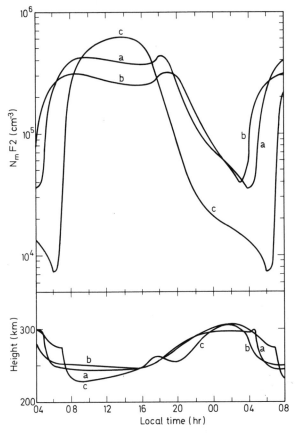

Fig. 77. Typical diurnal and seasonal variation of the peak density ($N_m F_2$) and height ($h_m F_2$) of the midlatitude F_2 region (Curves a, b, c refer to equinox, summer and winter, respectively). (After Strobel and McElroy [211])

of electron and ion temperature in the topside ionosphere is shown in Fig. 79. Another well-known F region anomaly is the *geomagnetic (Appleton) anomaly*. It consists of a minimum (trough) in the constant electron density contours over the dip equator and maxima (crests) at dip latitudes 10° to 20°. This anomaly extends into the topside iono-sphere where it gradually diminishes. The geomagnetic anomaly is primarily a daytime phenomenon and is absent near midnight [219]. The source of the anomaly is now understood to be upward plasma drift near the equator during daytime and field-aligned plasma diffusion; thermospheric winds are thought to be responsible for an observed asymmetry about the dip equator. The equatorial anomaly also exhibits a longitude and solar activity effect [220, 221].

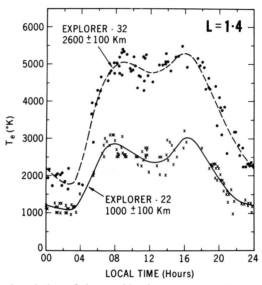

Fig. 78. Diurnal variation of the topside electron temperature at two altitudes showing a gradient $dT_e/dz \approx 1\,^\circ\mathrm{K/km}$ due to heat flux from the protonosphere. (Courtesy of L. H. Brace)

Another 'anomaly' is the fact that the *polar ionosphere* is maintained even in winter [222]. Photoionization can supply the F region as long as the solar zenith angle $\chi < 103^\circ$; thus, the maintenance of the ionosphere in winter is problematic for geographic latitudes $\varphi > 80^\circ$. Particle precipitation (in the energy range 10^2 to 10^3 eV) is expected to play an important role in maintaining the maximum electron density in winter of $N_m F_2 \cong 5 \times 10^4\,\mathrm{cm}^{-3}$ which requires an integrated production rate $(\int q\,dh)$ of the order $10^8\,\mathrm{cm}^{-2}\,\mathrm{sec}^{-1}$. Some of the observed features of the polar ionosphere may also be the result of thermospheric winds blowing across the polar caps (although at the dip pole, $I = 90^\circ$, the effects of winds should vanish), e.g., the universal time (UT) effect in the Antarctic according to which maxima of $h_m F_2$ and $N_m F_2$ occur near 0600 UT. A ring-like pattern of depletion in electron density (troughs) and enhancement in N (peaks) exist, somewhat asymmetrically, around the geomagnetic pole [223]. The region of density enhancement brackets the auroral oval which is associated with the precipitation of soft particles $E \sim \mathrm{keV}$ (see Fig. 80). Southward of the auroral zone, near 69° geomagnetic latitude, there is a sharp *through* in the F_2 region ionization [224] as well as a sharp decrease in the light ions (H^+, He^+) [225, 225a], phenomena which are associated with the *plasmapause* [133]. Its position is controlled by magnetic activity, but is also asymmetric with

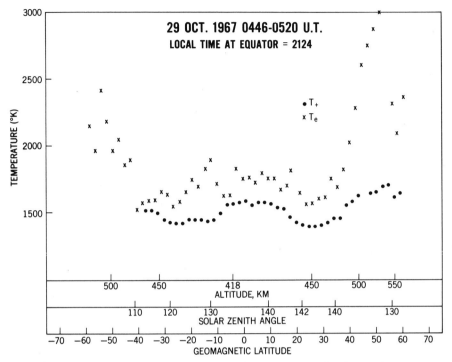

Fig. 79. Latitude variation of electron and ion temperature in the topside ionosphere obtained with a retarding potential analyzer experiment on OGO 4. (Courtesy of S. Chandra)

longitude (see Fig. 81). Over the polar cap there is a (supersonic) outflow
of light ions, the *polar wind*, which allows the escape of these ions into
the geomagnetic tail [134]. However, even at latitudes near the plasmapause where the geomagnetic flux tubes are not permanently open,
supersonic upward flow of light ions has been observed [226] (Fig. 82.)
The enhancement in ionization near 80° geomagnetic latitude may be
related to the neutral points of the magnetosphere; even the neutral
atmosphere appears to show similar effects at these latitudes. The
understanding of the interrelationship between high-latitude ionosphere,
neutral atmosphere and magnetospheric processes, however, is still in
its infancy.

Another feature of the ionosphere which is related to magnetospheric
processes is the *magnetic storm effect*. The F region shows increases
or decreases in electron density depending on the phase of the magnetic
storm, the latitude and the altitude. Storm-induced heating of the neutral
atmosphere affecting the atmospheric composition may be responsible for

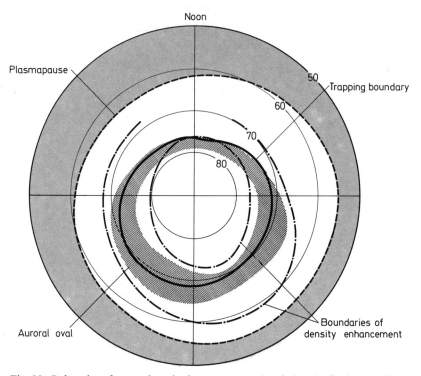

Fig. 80. Polar plot of auroral oval, plasmapause and polar peaks for intermediate
magnetic activity. (After Andrews and Thomas [223])

some of the observed features in the F region [227, 228, 228 a]. However,
magnetospheric compression and inflation (by ring-current plasma)
during the different phases of the storm may also affect the ionosphere
dynamically [229]. Although there exists a wealth of phenomenological
material of magnetic storm effects in the ionosphere, a full understanding
of all physical mechanisms operating in the ionosphere during magnetic
storms is still lacking [229 a].

A phenomenon associated with geomagnetic storms (though not
with all) is the *stable* (or *subauroral*) *red arc (SARARC)* which is identified by enhanced emission of the forbidden oxygen line at $\lambda = 6300 \text{\AA}$
[230]. It now appears, that the cause of the SARARC is heat conduction from the magnetosphere providing the excitation energy for the
optical emission ($E \sim 2$ eV), together with changes in the neutral atmosphere resulting from magnetic storm effects [231]. The combination
of these effects leads to a redistribution of charged and neutral particles,
an increase in T_e and T_i and a decrease in $N_m F_2$ causing an enhanced

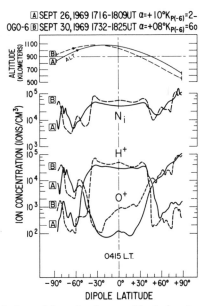

Fig. 81. Global variation of ions in the topside ionosphere showing the rapid decrease of light ions *(light ion trough)* associated with the plasmapause and its variation with geomagnetic activity. (After Taylor [225])

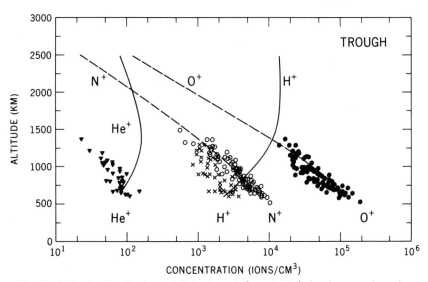

Fig. 82. Altitude distribution of light ions H^+ and He^+ in the trough region (plasmapause) showing evidence for maximum upward flux; heavy lines indicate diffusive equilibrium distributions; observations follow scale height of O^+ (and N^+) (cf. Fig. 30). (After [226])

emission of 6300Å. Since the emission intensity is an exponential function of T_e, a transition from relatively weak to bright emission is rapid when a certain threshold value of T_e is reached. This accounts for the fact that not all magnetic storms cause SARARC events which have an intensity of ~ 1 kR.

Another phenomenon occurring in the F region is *spread F* [232]. Its name is derived from the 'spread' appearance of ionogram traces due to diffuse reflections from irregularities in the electron density extending from below the F_2 peak into the topside ionosphere. While it is well established that spread F is associated with small-scale field-aligned irregularities in the electron density, an explanation for the origin of spread F is still lacking, although plasma instabilities are suspected. Furthermore, high latitude and equatorial spread F are most likely of completely different origin [233]. According to recent satellite observations, equatorial spread F seems to be associated with the presence of the minor meteoric ion Fe^+. This observation further increases the complexity of the spread F phenomenon [234].

The gross features of the terrestrial ionosphere are now reasonably well understood, primarily as the result of groundbased and spacecraft observations during the last decade. Many problems regarding the dynamics, energetics of the ionosphere and its interrelationship with the neutral atmosphere and the magnetosphere still remain. Another area where our understanding is still rather primitive is the ion-chemistry of the D region and the interaction between the lower atmosphere and the ionosphere.

IX.2. The Ionosphere of Mars

The Martian ionosphere was the first ionosphere of another planet whose existence was experimentally verified. The results from the Mariner 4 radio occultation experiment [235], however, showed a 'surprisingly underdeveloped' ionosphere relative to theoretical expectations, which in turn were based on terrestrial analogies [236, 236a]. Early explanations of the Mariner 4 observation were advanced in terms of E, F_1 and F_2 layers. It soon became apparent that a F_2 layer, i.e., one controlled by chemical processes *and* plasma diffusion is not consistent with the observation, since it would require an unacceptably low exospheric temperature. The low-lying 'underdeveloped' Martian ionosphere was recognized to be a photochemical equilibrium (Chapman) layer consisting of molecular ions (CO_2^+). This explanation raised the interesting question why CO_2 appears largely undissociated in spite of the high efficiency of photodissociation [49a]. Catalytic reactions with minor constituents together with strong turbulent mixing are presently thought to provide the best answer [8a].

Mariner 6 and 7 flybys provided additional electron density profiles of the daytime Martian ionosphere [237]; the nighttime ionosphere remains undetermined due to the lack of sensitivity of the S-band occultation experiment. Similarly, the termination of the Martian topside ionosphere and the nature of its boundary still remain to be observed. Fig. 83 shows the electron density profiles from the Mariner 4, 6 and 7

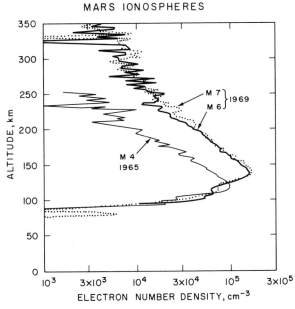

Fig. 83. Electron density profiles of the daytime Martian ionosphere obtained by the occultation experiments on Mariner 4, 6 and 7, representing solar minimum conditions (Mariner 4) and a period of intermediate solar activity (M 6 & 7). (After [237])

radio occultation data [237]. The Mariner 4 data, representing nearly solar minimum conditions show a peak density $N_m = 9 \pm 1 \times 10^4 \, \text{cm}^{-3}$ at an altitude $h_m \simeq 120 \, \text{km}$, whereas Mariner 6 and 7 observations during a period of intermediate solar activity, four years later, show an increased peak density of 1.6 and $1.7 \times 10^5 \, \text{cm}^{-3}$ at an altitude $h_m \simeq 135 \, \text{km}$. In addition to the main layer of ionization, which is now generally accepted to be an F_1-layer, a secondary electron density ledge was observed $\sim 25 \, \text{km}$ below the peak, by Mariners 4, 6 and 7. This layer appears to be due to ionization by soft X-rays and therefore corresponds to an E layer. A substantial D-region, although not yet detected, is expected to extend almost down to the Martian surface, where the pressure is ~5 mb [239, 240].

Interpretation of the scale height with which the plasma density decreases above the peak, leads to exospheric temperatures $T \cong 275°$K for Mariner 4; $T \cong 388° \pm 54°$K for Mariner 6 and $T \cong 425° \pm 35°$K for Mariner 7 observations. The low exospheric temperature for Mariner 4 seems to be associated with the solar minimum condition. An alternative interpretation assumed a much higher exospheric temperature and interpreted the observed plasma scale height as being depressed (by a factor of 2) as the result of a soft solar wind interaction with the Martian ionosphere [153]. However, such an explanation is not consistent with the Mariner 6, 7 and 9 observations [237, 238], since the observed plasma scale heights is $H_N \cong 2H_n$, corresponding to the scale height of an unperturbed photochemical equilibrium (Chapman) layer. Fig. 84 shows a daytime electron density profile obtained from the

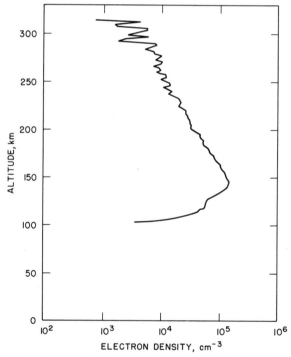

Fig. 84. Daytime electron density profile of the Martian ionosphere obtained from the Mariner 9 radio occultation experiment. (After [238])

Mariner 9 radio occultation experiment [238]. Mariner 9, an orbiting spacecraft, makes repeated observations of the Martian ionosphere possible. The shape of the electron density profile seems to remain quite constant with a scale height $H_N \cong 38$ km, in good agreement with the scale height for the neutral atmosphere determined from UV airglow observations, according to which $H_n \cong 19$ km [241]. The height of the ionization peak $h_m \cong 140$ km is somewhat higher than that of the previously measured profiles. Variations of h_m and N_m with solar zenith angle χ illustrated in Figs. 85 and 86 show clearly the behavior characteristic of a Chapman layer ($h_m \propto \ln \cdot \sec \chi$ and $N_m \propto \cos^{\frac{1}{2}} \chi$) and lend further credence to the interpretation of the Martian ionosphere as an F_1 layer. (For larger zenith angles there seems to be a departure from a simple Chapman layer behavior) [242.]

Recent atmospheric models consistent with the Mariner observations assume CO_2 as the predominant constituent, with N_2, O, CO and O_2 as minor constituents [243, 244]. The Martian ionosphere is thought to be produced by photoionization of CO_2 followed by the competing reactions

$$CO_2^+ + O \rightarrow O_2^+ + CO \qquad (9\text{–}2)$$

and

$$CO_2^+ + e \rightarrow CO + O \qquad (9\text{–}3)$$

leading to O_2^+ as the predominant ionic constituent, since the charge

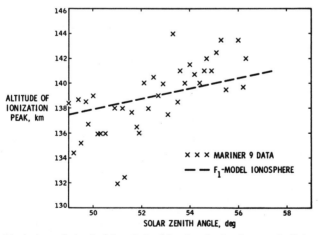

Fig. 85. Variation of the height of the Martian ionization peak (h_m) with solar zenith angle χ from Mariner 9 occultation observations, exhibiting Chapman layer behavior ($h_m \propto \ln \sec \chi$). (Courtesy of A. Kliore and S. I. Rasool)

---

* Similar observations were made with the USSR Mars-2 orbiter (M. A. Kolosov et al., Radiotekhnika i Elektronika 12, 2483—2490, 1972).

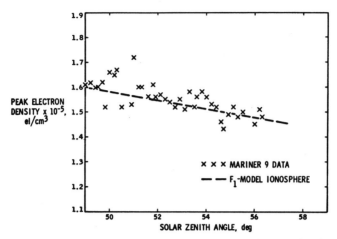

Fig. 86. Variation of the peak electron density N_m in the Martian ionosphere observed by the Mariner 9 occultation experiment as function of solar zenith angle, exhibiting Chapman layer behavior $(N_m \propto \cos^{\frac{1}{2}}\chi)$. (Courtesy of A. Kliore and S. I. Rasool)

transfer process is more efficient below 250 km. Escape of O, CO and N can take place as the result of the exothermic dissociative recombination reactions, which provide sufficient energy for the escape of these constituents [42] (cf. Fig. 17) and since the exobase is at an altitude of $\lesssim 200$ km.

At present, there are still problems relating to the explanation of the Martian ionosphere in terms of the recently favored lower XUV fluxes, together with the resulting thermal structure of the Martian upper atmosphere [245, 246]. There seems to be a need for an additional ionization source, such as solar wind protons, to explain the observed electron densities [246]. Furthermore, experimental clues regarding the type of interaction of the solar wind with the Martian ionosphere are still lacking (although it now appears that this interaction is indirect [151b]), as is information on ion composition and charged particle temperatures.

IX.3. The Ionosphere of Venus

In addition to the S-band, Mariner 5 also carried a dual (lower) frequency occultation experiment (423.3 and 49.8 MHz) which has a substantially higher sensitivity [247]. This experiment allowed the detection of both, the dayside and nightside ionosphere of Venus and gave the indication of a sharp boundary of the dayside ionosphere (ionopause or plasmapause), thought to be the result of the solar wind interaction with the ionospheric plasma due to the absence of a significant planetary magnetic field. The nightside ionosphere, having a much lower electron density, appears to extend to much greater distances in the antisolar direction, consistent with a direct interaction of the solar wind with the topside ionosphere of Venus. Fig. 87 shows the day- and nightside profiles of the Venus ionosphere obtained by the dual frequency occultation experiment. The boundary between the solar wind and the ionosphere, i.e., the ionopause, based on the pressure balance of solar wind and ionosphere plasma at the dayside observation point, has a configuration shown in Fig. 50.

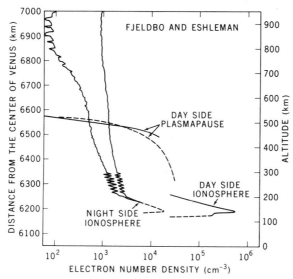

Fig. 87. Dayside and nightside electron density profile of the Venus ionosphere obtained by the Mariner 5 dual frequency occultation experiment, indicating the termination of the topside ionosphere by solar wind interaction. (After [247])

The main layer of the dayside ionosphere having a peak at an altitude $h_m \cong 140$ km with a density $N_m \cong 5 \times 10^5$ cm^{-3} can be explained as a CO_2^+ photochemical equilibrium layer [248, 249]. The large scale height in the topside ionosphere suggests the presence of light ions at relatively high plasma temperature $(T_e + T_i)$ under diffusive control [250]. A model of the dayside ionosphere [95] which is able to reproduce the observed features reasonably well is shown in Fig. 88. This model assumes He$^+$ as major ion in the topside ionosphere, above the region where CO_2^+ is dominant.

Although the presence of helium on Venus is still conjectural, whereas hydrogen has been observed both on Mariner 5 and Venera 4 [251, 252],

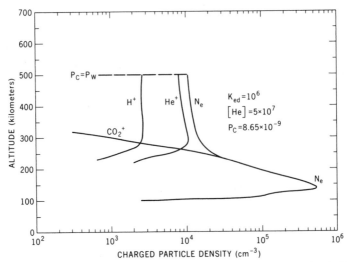

Fig. 88. Theoretical profile of the dayside ionosphere of Venus showing individual ion species, obtained for boundary conditions appropriate to the Mariner 5 observations. (After [95])

the required density of He at 500 km, $n(\text{He}) = 5 \times 10^7 \text{ cm}^{-3}$ is quite reasonable on the basis of an earthlike crust for Venus and similar outgassing rate of He, produced by radioactive decay of Uranium and Thorium*. The Venus ionosphere model shown in Fig. 88 was obtained from the simultaneous solution of the momentum, chemical and heat balance equations of the electron, ion and neutral gases, with the lower boundary condition based on models and the upper boundary condition satisfying the balance between solar wind streaming pressure and ionospheric plasma pressure $N\mathscr{k}(T_e + T_i)$ [95]. A small horizontal magnetic field in the ionosphere of Venus of $\lesssim 10$ gamma (not inconsistent with observational limits) suppressing vertical heat conduction leads to the plasma temperatures (shown in Fig. 33) which are required for pressure balance with the solar wind. Heating originating from the solar wind interaction, although probably small compared to that from EUV absorption, provides a means for stabilizing the topside ionosphere. Without a solar-wind related heat source, small changes in the solar wind velocity and density could cause large excursion of the ionopause level. With an efficiency of 1 to 5% for the solar wind heating, the ionopause could remain stable for the expected range of solar wind

---

* Gamma-ray spectrometric analysis performed with the USSR Venera-8 probe showed that the content of natural radioactive elements in rocks on the Venus surface is similar to that of igneous rocks on Earth.

pressures. This particular model, as well as others, is of course speculative due to the absence of many important parameters which will have to come from future measurements.

Another model of the topside ionosphere of Venus [253] based on upward flowing protons due to the large polarization field of H^+ in a CO_2^+ ionosphere (cf. 5–20), suffers from the fact that the required density of H^+, $n(H^+) \cong 10^4$ cm^{-3} cannot be obtained with the neutral hydrogen densities observed by the UV airglow experiments. Furthermore, a topside ionosphere consisting of H^+ would be subject to the Rayleigh-Taylor instability; a stable ionopause configuration requires that the mean ion mass $m_i > 1.7$ AMU [170]. Another possible constituent of the topside ionosphere of Venus would be O^+ at high plasma temperatures, provided there is sufficient neutral oxygen in the Venus upper atmosphere; UV airglow observations on the forthcoming Mariner Venus/Mercury flyby mission in 1973 should be able to shed some light an these alternatives.

The exospheric temperature of Venus at the time of the Mariner 5 mission was found to be $T \cong 650°$K, whereas the inferred plasma temperatures for pressure balance with the solar wind, are of the order $(T_e + T_i) = 5300°$K, implying the absence of thermal equilibrium in the topside ionosphere of Venus. The observation of a nightside ionosphere represents an interesting problem in view of the fact, that the solar day on Venus is ~ 116 days, and thus it is impossible to maintain an ionospheric layer as observed $(N_m \cong 10^4$ cm$^{-3})$ without a source other than solar EUV during the long Venus night. A proposal has been made [254] according to which the dayside topside ionosphere helps to maintain the nightside by means of transport of light ions (He^+) which charge exchange with CO_2. The CO_2^+ ions formed in this fashion could diffuse downward and maintain the nightside ionosphere. The horizontal transport velocity required for He^+ is $v \sim 100$ m sec^{-1}. (It is interesting to note that this corresponds to the four-day superrotation which has been observed at lower atmospheric levels; this atmospheric superrotation may also alleviate some of the problems of maintaining the nightside ionosphere.) The transport model for maintaining the nightside ionosphere, however, suffers from the fact, that downward diffusion would tend to form the peak of the nightside layer at an altitude greater than that of the dayside photochemical equilibrium layer, whereas the observations seem to indicate that h_m is about the same on the day and nightside. Another possibility for maintaining the nightside ionosphere is leakage of solar wind protons into the nightside atmosphere [255]. The solar wind protons can charge-exchange with CO_2, producing hot hydrogen atoms which penetrate deeper into the atmosphere, scattering and ionizing CO_2 on the way. A hot hydrogen atom suffers about 20 collisions before thermalizing, heating and ionizing the atmosphere in its path. Although on the dayside the ionization and heating due to leaking solar wind protons may be small compared to that due to the solar EUV radiation, the solar wind may represent the primary ionization source for the nightside. Only about 1 to 2% of the solar wind energy flux is required to produce the observed nightside electron density peak of $\sim 10^4$ cm^{-3} at the appropriate height (Fig. 89). The nightside topside ionosphere, however, may possibly be the result of transport of light ions from the dayside.

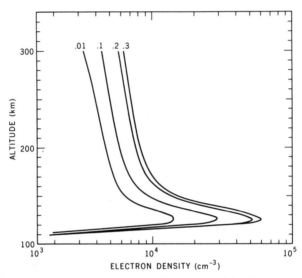

Fig. 89. Theoretical electron density profiles of the nightside Venus ionosphere resulting from ionization by solar wind protons leaking into the nightside "tail". Fraction of solar wind energy flux responsible for ionization is indicated. (Courtesy of R. E. Hartle and J. R. Herman)

As in the case of Mars, the presence of atomic oxygen in the atmosphere of Venus may lead to O_2^+ as the predominant ion in the main photochemical F_1 layer. There is a similar problem regarding the adequacy of the XUV fluxes for ionization and heating for Venus as discussed for Mars [246]. The observed electron density profile of the dayside ionosphere of Venus also shows a subsidiary layer, about 15 km below the F_1 peak, which seems to be produced by soft X-rays and therefore represents an E-layer. It has also been speculated that in the Venus D-region, cluster ions of the type $CO_2^+ \cdot (CO)_n$ should be of great importance, in fact, they may be associated with the Venus UV-haze layer, since cluster ions can represent micron-size particles [256].

For the ionosphere of Venus a single dayside and a single nightside electron density profile is available from the Mariner 5 observations, also giving us a first glimpse of the solar wind interaction with a planet distinctly different from that with Earth or the moon. Theoretical explanations based on this limited amount of experimental data must remain highly speculative until further observations, most likely during this decade, will provide the necessary input data to advance our present understanding of the ionosphere of Venus.

IX.4. The Ionosphere of Jupiter

Although no experimental data is yet available on the ionosphere of Jupiter, there is no doubt about its existence since the radio emissions from this planet are obviously associated with an ionosphere and magnetosphere [257]. There exist several speculative models of the Jovian ionosphere [258, 259, 260]. They all assume an ionosphere in photochemical equilibrium with a maximum density $10^5 \lesssim N_m(\mathrm{cm}^{-3})$ $\lesssim 10^6 \star$ at an altitude ~ 300 km above the cloud top level. The major ion is supposed to be H^+ or possibly H_3^+, arising from various charge transfer reactions in the predominantly H_2/He Jovian upper atmosphere. On Jupiter as on Earth the ionospheric plasma also represents an integral part of the magnetosphere [150, 151].

Based on our experience with Mars and Venus, it would be highly surprising if the first experimental data on the Jovian ionosphere would agree with our present models. With planned space probe missions to the outer planets, particularly to Jupiter and Saturn, more detailed experimental information on the Jovian ionosphere will hopefully become available during this decade, although the first data can be expected from the S-band occultation experiment with Pioneer 10 which is due to arrive near Jupiter in the winter of 1973/74.

$\star$ It is a curious coincidence that the peak densities of the ionospheres of Venus, Earth, Mars and Jupiter are virtually the same, in spite of the fact that these planets differ widely in their size, distance from the sun and atmospheric structure.

Appendix. Physical Data for the Planets and their Atmospheres

Table A.1. *Planetary Data**

	Mean Radius (km)	Mean Density (g cm^{-3})	Average Distance from Sun 10^6 km	AU	Length of Year (Days)	Period of Rotation (d)	Inclination of Equator to Orbit Plane ($\delta°$)
(Inner Planets)							
Mercury	2439	5.42	58	0.39	88	58.7	<28
Venus	6050	5.25	108	0.72	225	-243**	<3
Earth	6371	5.51	150	1.00	365	1.00	23.5
Mars	3390	3.96	228	1.52	687	1.03	25
(Outer Planets)							
Jupiter	69500	1.35	778	5.2	4330	0.41	3.1
Saturn	58100	0.69	1430	9.5	10800	0.43	26.7
Uranus	24500	1.44	2870	20	30700	-0.89	98.0
Neptune	24600	1.65	4500	30	60200	0.53	28.8
Pluto	?	?	5900	39	90700	(6.39)	?

* For gravitational acceleration and escape velocity see Tables 1 and 5 of text.
** Minus sign denotes retrograde motion.

Table A.2. *Composition of Planetary Atmospheres*
1) Terrestrial Planets

Percent by Concentration

	Lower Atmosphere (mixing region)								Upper Atmosphere (ionization peak)			
	CO_2	N_2,A	O_2	H_2O	HCl	HF	CO	NH_3	O	H	He	CO_2
Venus	~97	<3	$<4 \times 10^{-3}$	10^{-2}	10^{-4}	2×10^{-6}	2×10^{-2}	≤ 0.1	≤ 1	10^{-4}	$(10^{-1}?)$	≤ 98
Earth	0.3	78, <1	21	≤ 1			10^{-5}		98.9	0.1	1	
Mars	~98	<2	10^{-1}	$\leq 10^{-1}$			10^{-1}		0.5—1	10^{-4}	(?)	≤ 98

2) Outer Planets

Total content (cm^{-2}) above clouds

	H_2	He	CH_4	NH_3
Jupiter	1.8×10^{26}	$<9 \times 10^{25}$	1.2×10^{23}	2.6×10^{22}
Saturn	3.7×10^{26}		9.4×10^{23}	$<6.7 \times 10^{21}$
Uranus	1.3×10^{27}		9.4×10^{23}	
Neptune			1.6×10^{25}	

References

1. Kockarts, G.: Mean molecular mass and scale heights of the upper atmosphere. Ann. Geophys. **22**, 91 (1966).
2. Nicolet, M.: The structure of the upper atmosphere. In: Odishaw, H. (ed.), Research in Geophysics, Vol. I, pp. 243—275. Cambridge: The MIT Press 1964.
3. Mange, P.: Diffusion in the thermosphere. Ann. Geophys. **17**, 277 (1961).
4. Chapman, S., Cowling, T. G.: The Mathematical Theory of Nonuniform Gases. Cambridge University Press, second ed., 431 pp. 1952.
5. McElroy, M. B., Hunten, D. M.: The ratio of deuterium to hydrogen in the Venus atmosphere. J. Geophys. Res. **74**, 1720—1739 (1969).
6. Hines, C. O.: Eddy diffusion coefficients due to instabilities in internal gravity waves. J. Geophys. Res. **75**, 3937—3939 (1970).
7. Colegrove, F. D., Hanson, W. B., Johnson, F. S.: Eddy diffusion and oxygen transport in the lower thermosphere. J. Geophys. Res. **40**, 4931—4942 (1965).
8. Shimazaki, T.: Effective eddy diffusion coefficient and atmospheric composition in the lower atmosphere. J. Atmospheric Terrest. Phys. **33**, 1383—1401 (1971).
8a. McElroy, M. B., Donahue, T. M.: Stability of the Martian atmosphere. Science **177**, 986—988 (1972).
9. Donahue, T. M.: Deuterium in the upper atmosphere of Venus and earth. J. Geophys. Res. **74**, 1128—1137 (1969).
10. Hunten, P. M.: Hydrogen isotopes around the planets. Comm. Astrophys. & Space Phys. **III/1**, 1—5 (1971).
11. Izakov, M. N.: On theoretical models of the structure and dynamics of the earth's thermosphere. Space Sci. Rev. **12**, 261—298 (1971).
12. Hines, C. O.: Internal gravity waves at ionospheric heights. Can. J. Phys. **38**, 1441—1481 (1960).
13. Charney, J. G., Drazin, P. G.: Propagation of planetary-scale disturbances from the lower into the upper atmosphere. J. Geophys. Res. **66**, 83—109 (1961).
14. Bates, D. R.: Some problems concerning the terrestrial atmosphere above about 100 km level. Proc. Roy. Soc. A **253**, 451—462 (1959).
15. Stewart, R. W.: Radiative terms in the thermal conduction equation for planetary atmospheres. J. Atmospheric Sci. **25**, 744—749 (1968).
16. Harris, I., Priester, W.: Time dependent structure of the upper atmosphere. J. Atmospheric Sci. **18**, 286—301 (1962).
17. Rishbeth, H., Moffett, R. J., Bailey, G. J.: Continuity of air motion in the mid latitude thermosphere. J. Atmospheric Terrest. Phys. **31**, 1035—1047 (1969).

18. Jacchia, L. G.: Atmospheric density variations during solar maximum and minimum. Paper 17, Annls. IQSY, Vol. **5** Solar-Terrestrial Physics: Terrestrial Aspects pp. 333—339, MIT Press 1969.
19. Dickinson, R.E., Pablo Lagos, C., Newell, R. E.: Dynamics of the neutral gas in the thermosphere for small Rossby number motions. J. Geophys. Res. **73** (13), 4299—4313 (1968). (Correction: J. Geophys. Res. **73** (21), 6876 (1968).
20. Volland, H., Mayr, H. G.: A theory of the diurnal variations of the thermosphere. Ann. Geophys. **25**, 907—919 (1970).
21. Chandra, S., Stubbe, P.: The diurnal phase anomaly in the upper atmospheric density and temperature. Planetary Space Sci. **18**, 1021—1033 (1970).
21a. Mayr, H. G., Volland, H.: Diffusion model for the phase delay between thermospheric density and temperature. J. Geophys. Res. **77**, 2359—2367 (1972).
22. Bates, D. R., Moffett, R. J.: Response of a planetary thermosphere to heating by solar radiation. Planetary Space Sci. **16**, 1531—1537 (1968).
23. Volland, H., Mayr, H. G.: Response of the thermospheric density to auroral heating during geomagnetic disturbances. J. Geophys. Res. **76**, 3764—3776 (1971).
24. Timothy, A. F., Timothy, J. G.: Long term intensity variations in the solar He II Lyman-alpha line. J. Geophys. Res. **75**, 6950—6958 (1970).
25. Hall, L. A., Hinteregger, H. E.: Solar radiation in the extreme ultraviolet and its variation with solar rotation. J. Geophys. Res. **75**, 6959—6965 (1970).
26. Bauer, S. J.: Solar cycle variation of planetary exospheric temperatures. Nature **232**, 101—102 (1971).
27. Tsederberg, N. V.: Thermal Conductivities of Gases and Liquids. MIT Press 1965.
28. Dalgarno, A., Smith, F. J.: Planetary Space Sci. **9**, 1 (1962).
    Dalgarno, A., Smith, F. J.: Proc. Roy. Soc. A **267**, 417 (1962).
29. Stewart, R. W., Hogan, J. S.: Solar cycle variation of exospheric temperatures on Mars and Venus: A prediction for Mariner 6 and 7. Science **165**, 386—388 (1969).
30. Matora, I. M.: Exosphere temperatures of the Earth, Mercury, Venus, Mars, Jupiter and the solar corona. Soviet Phys. Doklady (Astron.) **15**, 83 (1970).
31. Jeans, Sir James: The Dynamical Theory of Gases. Cambridge: University Press; (4th edition, Dover Publications, N.Y.) 1925.
32. Biutner, E. K.: The dissipation of gas from planetary atmospheres. Soviet Astron. AJ **2**, 528—537 (1958).
33. Öpik, E. J., Singer, S. F.: Distribution of density in a planetary exosphere II. Phys. Fluids **4**, 221—233 (1961).
34. Chamberlain, J. W.: Planetary coronae and atmospheric evaporation, Planetary Space Sci. **11**, 901—96 (1963).
35. Johnson, F. S.: Density of an exosphere. Ann. Geophys. **22**, 56—61 (1966).
35a. Brinton, H. C., Mayr, H. G.: Temporal variations of thermospheric hydrogen derived from in-situ measurements. J. Geophys. Res. **76**, 6198 (1971).
36. Chamberlain, J. W., Campbell, F. J.: Rate of evaporation of a non-Maxwellian atmosphere. Astrophys. J. **149**, 687—705 (1967).
37. Brinkman, R. T.: Departures from Jeans' escape rate for H and He in the earth's atmosphere. Planetary Space Sci. **18**, 449—478 (1970).
38. Öpik, E. J.: Selective escape of gases. Geophys. J. **7**, 490 (1963).
39. Hagenbuch, K. M., Hartle, R. E.: Simple model for a rotating neutral planetary exosphere. Phys. Fluids **12**, 1551 (1969).

40. Hartle, R. E.: Model for rotating and non-uniform planetary exospheres. Phys. Fluids **14**, 2592—2598 (1971).

40a. Hartle, R. E., Ogilvie, K. W., Wu, C. S.: Neutral and ion-exospheres in the solar wind. Planetary Space Sci. **21** (1973).

41. Brinkmann, R. T.: Mars: Has nitrogen escaped? Science **174**, 944 (1971).

42. McElroy, M. B.: Mars: An evolving atmosphere. Science **175**, 443—445 (1972).

43. Sagan, C.: Origins of the atmospheres of earth and planets, Intern. Dict. Geophys., S. K. Runcorn, ed. pp. 97—104, London: Pergamon Press 1968.

44. Johnson, F. S.: Origin of planetary atmospheres. Space Sci. Rev. **9**, 303—324 (1969).

45. Banks, P. M., Johnson, H. E., Axford, W. I.: The atmosphere of Mercury. Comm. Astrophys. & Space Phys. **II/6**, 214—220 (1970).

46. Rasool, S. I., de Bergh, C.: The runaway greenhouse and the accumulation of CO_2 in the Venus atmosphere. Nature **226**, 1037—1039 (1970).

47. Ingersoll, A. P., Leovy, C. B.: The atmospheres of Mars and Venus. Ann. Rev. Astronomy Astrophys. **9**, 147—182 (1971).

48. Hunten, D. M.: Composition and structure of planetary atmospheres. Space Sci. Rev. **12**, 539—599 (1971).

48a. Lewis, J. S.: Metal/silicate fractionation in the solar system. Earth Planet. Sci. Lett. **15**, 286—290 (1972).

49. Donahue, T. M.: Aeronomy of CO_2 atmospheres: A review. J. Atmospheric Sci. **28**, 895—900 (1971).

49a. Hunten, D. M.: Aeronomy of CO_2 atmospheres, Comm. Astrophys. & Space Phys. **4**, 1—5 (1972).

50. McElroy, M. B.: Atmospheric Composition of the Jovian Planets, J. Atmospheric Sci. **26**, 748—812 (1969).

51. Trafton, L. M., Münch, G.: The structure of the atmospheres of the major planets. J. Atmospheric Sci. **26**, 813—826 (1969).

52. de Jager, C.: Solar ultraviolet and x-ray radiation. Ch. 1 (pp. 1—42) Research in Geophys. Vol. **1** (Sun, Upper Atmosphere and Space) H. Odishaw, ed., MIT Press 1964.

52a. Ivanov-Kholodnyi, G. S., Nikolski, G. M.: The Sun and the Ionosphere (short wave solar radiation and its effect on the ionosphere) Nauka, Moscow, 1969, Engl. Transl. NASA TTF-654, 1972.

53. Hinteregger, H. E., Hall, L. A., Schmidke, G.: Solar XUV radiation and neutral particle distribution in July 1963. Space Research **V**, edited by D. G. King-Hele, P. Miller, G. Righini, 1175—1190, Amsterdam: North-Holland, 1965.

54. Hall, L. A., Higgins, J. E., Chagnon, C. W., Hinteregger, H. E.: Solar-cycle variation of the extreme ultraviolet radiation. J. Geophys. Res. **74**, 4181—4183 (1969).

55. Hinteregger, H. E.: The extreme solar ultraviolet spectrum and its variation during a solar cycle. Ann. Geophys. **26**, 547—554 (1970).

56. Chamberlain, J. W.: Physics of the Aurora and Airglow. N. Y.: Academic Press 1961.

57. Dalgarno, A., Degges, T. C.: CO_2^+ dayglow on Mars and Venus. Planetary Atmospheres, pp. 337—345, Sagan et al. (eds), I.A.U., 1971.

58. Noxon, J. F.: Day airglow. Space Sci. Rev. **8**, 92—134 (1968).

59. Barth, C. A.: The ultraviolet spectroscopy of the planets, Ch. 10 (pp. 177—218). The Middle Ultraviolet: Its Science and Technology, A. E. S. Green, Ed., John Wiley Publ. 1966.

60. Barth, C. A.: Planetary ultraviolet spectroscopy. Appl. Opt. **8**, 1295—1304 (1969).

61. Aikin, A. C.: X-ray glow from planetary atmospheres. Nature **227**, 1334 (1970).

62. Chapman, S.: The absorption and dissociative or ionizing effect of mono-chromatic radiation in an atmosphere on a rotating earth. Proc. Phys. Soc. (London) **43**, 26—45; The absorption and dissociative or ionizing effect of monochromatic radiation in an atmosphere on a rotating earth. II. Grazing incidence. Proc. Phys. Soc. (London) **43**, 483—501 (1931).

63. McDaniel, E. W.: Collision Phonomena in Ionized Gases, N. Y.: John Wiley and Sons 1964.

64. Hudson, R. D.: Critical review of ultraviolet photo-absorption cross sections for molecules of astrophysical and aeronomic interest. Rev. Geophys. **9**, 305—406 (1971).

65. Swider, W., Gardner, M. E.: On the accuracy of certain approximations for the Chapman function. Environmental Research Papers No. 272, Bedford, Mass.: Air Force Cambridge Research Labs 1967.

66. Stewart, A. I.: Photoionization coefficients and photoelectron impact excitation efficiencies in the daytime ionosphere. J. Geophys. Res. **75**, 6333—6338 (1970).

67. Van Allen, J. A.: On the nature and intensity of cosmic radiation. Physics & Medicine of the Upper Atmosphere and Space, White and Benson, ed., Albuquerque: Univ. of New Mexico Press 1952.

68. Webber, W.: The production of free electrons in the ionospheric D layer by solar and galactic cosmic rays and the resultant absorption of radio waves. J. Geophys. Res. **67**, 5091—5106 (1962).

69. Rao, U. R.: Solar modulation of galactic cosmic radiation. Space Sci. Rev. **12**, 719—809 (1972).

70. Dubach, J., Barker, W. A.: Charged particle induced ionization rates in planetary atmospheres. J. Atmospheric Terrest. Phys. **33**, 1287—1288 (1971).

71. Dalgarno, A.: Range and energy loss. Ch. 15 (pp. 622—642) Atomic and Molecular Processes, D. R. Bates (ed.), N. Y.: Academic Press 1962.

72. Singer, S. F., Maeda, K.: Energy dissipation of spiraling particles in the polar atmospheres. Arkiv Geofysik **3**, 531—538 (1961).

73. Berger, M. J., Seltzer, S. M., Maeda, K.: Energy deposition by auroral electrons in the atmosphere. J. Atmospheric Terrest. Phys. **32**, 1015—1045 (1970).

74. Brown, R. R.: Solar cosmic ray effects in the lower ionosphere of Venus. Planetary Space Sci. **17**, 1923—1926 (1969).

75. McKinley, D. W. R.: Meteor Science and Engineering, New York: McGraw-Hill 1961.

76. Lebedinets, V. N., Shushkova, V. B.: Meteor ionization in the E layer. Planetary Space Sci. **18**, 1659—1663 (1970).

77. Henry, R. J. W., McElroy, M. B.: Photoelectrons in planetary atmospheres. The Atmophere of Venus and Mars, J. C. Brandt and M. B. McElroy, ed., New York: Gordon & Breach, Science Publishers 1968.

78. Spitzer, Lyman: Physics of Fully Ionized Gases. New York: Interscience Publ., 105 pp. 1956.

79. Hanson, W. B., Johnson, F. S.: Electron temperatures in the ionosphere. Mem. Soc. Roy. Sci. Liege **4**, 390—423 (1961).

80. Hanson, W. B.: Electron temperatures in the ionosphere. Space Res. **III**, 282—303 (1963).

81. Dalgarno, A., McElroy, M. B., Moffett, R. J.: Electron temperatures in the ionosphere. Planetary Space Sci. **11**, 463—484 (1963).
82. Banks, P. M.: The thermal structure of the ionosphere. Proc. IEEE **57**, 258—281 (1969).
83. Nisbet, J. S.: Photoelectron escape from the ionosphere. J. Atmospheric Terrest. Phys. **30**, 1257—1278 (1968).
84. Nagy, A. F., Banks, P. M.: Photoelectron fluxes in the ionosphere, J. Geophys. Res. **75**, 6260—6270 (1970).
85. Schunk, R. W., Hays, P. B.: Photoelectron energy losses to thermal electrons. Planetary Space Sci. **19**, 113 (1971).
86. Takayanagi, J., Itikawa, Y.: Elementary processes involving electrons in the ionosphere. Space Sci. Rev. **11**, 380—450 (1970).
87. Schunk, R. W., Hays, P. B., Itikawa, Y.: Energy loss of low energy photoelectrons to thermal electrons. Planetary Space Sci. **19**, 125—126 (1971).
88. Swartz, W. E., Nisbet, J. S., Green, A. E. S.: Analytic expression for energy transfer rate from photoelectrons to thermal electrons. J. Geophys. Res. **76**, 8425—8426 (1971).
89. Stubbe, P.: Energy exchange and thermal balance problems. Proc. Sympos. "Ionosphere-Magnetosphere Interactions". J. Sci. Ind. Res. (India) **30**, 379—387 (1971).
89a. Dalgarno, A.: Inelastic collisions at low energies, Can. J. Chem. **47**, 1723—1729 (1969).
90. Stubbe, P.: Frictional forces and collision frequencies between moving ion and neutral gases. J. Atmospheric Terrest. Phys. **30**, 1965—1985 (1968).
91. Mayr, H. G., Volland, H.: Model of the magnetospheric temperature distribution. J. Geophys. Res. **73**, 4851 (1968).
92. Cole, K. D.: Joule heating of the upper atmosphere. Australian J. Phys. **15**, 223—235 (1962).
93. Cole, K. D.: Electrodynamic heating and movement of the thermosphere. Planetary Space Sci. **19**, 59—75 (1971).
94. Rees, M. H., Walker, J. C. G.: Ion and electron heating by auroral electric fields. Ann. Geophys. **24**, 193 (1968).
95. Herman, J. R., Hartle, R. E., Bauer, S. J.: The Dayside Ionosphere of Venus. Planetary Space Sci. **19**, 443—460 (1970).
96. Hasted, J. B.: Physics of Atomic Collisions. Washington: Butterworth 1964.
97. Bates, D. R.: Reactions in the ionosphere. Contemp. Phys. **11**, 105—124 (1970).
98. Bates, D. R., Dalgarno: Electronic recombination. Ch. 7 (pp. 245—271). Atomic and Molecular Processes, D. R. Bates (ed.), New York: Academic Press 1962.
99. Biondi, M. A.: Atmospheric electron-ion and ion-ion recombination processes. Can. J. Chem. **47**, 1711 (1969).
100. McDaniel, E. W., Cermak, V., Dalgarno, A., Ferguson, E. E., Friedman, L.: Ion Molecule Reactions, New York: Wiley-Interscience 1970.
101. Polyanyi, J. C.: Chemical processes. Ch. 21. (pp. 807—855). Atomic and Molecular Processes, D. R. Bates, ed., New York: Academic Press 1962.
102. Rapp, D.: Acidentally resonant asymmetric charge exchange in the protonosphere. J. Geophys. Res. **68**, 1773—1775 (1963).
103. Dungey, J. W.: The electrodynamics of the outer ionosphere. The Physics of the Ionosphere, p. 229, London: Phys. Soc. 1955.
104. Bates, D. R., Patterson, T. N. L.: Hydrogen atoms and ions in the thermosphere and exosphere. Planetary Space Sci. **5**, 257—273 (1961).

105. Hanson, W. B., Patterson, T. N. L., Degaonkar, S.: Some deductions from a measurement of the hydrogen ion distribution in the high atmosphere. J. Geophys. Res. **68**, 6203—6205 (1963).

106. Fehsenfeld, F. C., Ferguson, E. E.: Thermal energy reaction rate constants for H^+ and CO^+ with O and NO. J. Chem. Phys. **56**, 3066—3070 (1972).

107. Ferguson, E. E.: Ionospheric ion-molecule reaction rates. Rev. Geophys. **5**, 305—327 (1967).

108. Ferguson, E. E.: D-region ion chemistry, Rev. Geophys. **9**, 997—1008 (1971).

109. Branscomb, L.: A review of photodetachment and related negative ion processes relevant to aeronomy. Ann. Geophys. **29**, 49—66 (1964).

110. Phelps, A. V.: Laboratory studies of attachment and detachment processes of aeronomic interest. Can. J. Chem. **47**, 1783—1793 (1969).

111. Peterson, V. L., Van Zandt, T. E., Norton, R. B.: F-region night-glow emissions of atomic oxygen, 1. Theory. J. Geophys. Res. **71**, 2255 (1966).

112. Hanson, W. B.: Radiative recombination of atomic oxgen ions in the nighttime F region. J. Geophys. Res. **74**, 3720—3722 (1969).

113. Hanson, W. B.: A comparison of the oxygen ion-ion neutralization and radiative recombination for producing the ultraviolet nightglow. J. Geophys. Res. **75**, 4343—4346 (1970).

114. Ferraro, V. C. A.: Diffusion of ions in the ionosphere. Terr. Magn. Atmos. Electr. **50**, 215—222 (1945).

115. Kendall, P. C., Pickering, W. M.: Magnetoplasma diffusion at F2-region altitudes. Planetary Space Sci. **15**, 825—833 (1967).

116. Bauer, S. J.: Diffusive equilibrium in the topside ionosphere. Proc. IEEE **57**, 1114—1118 (1969).

117. Mange, P.: The distribution of minor ions in electrostatic equilibrium in the high atmosphere. J. Geophys. Res. **65**, 3833 (1960).

118. Eddington, Sir A. S.: The Internal Constitution of the Stars. N. Y.: Dover Publications 1959.

119. Bauer, S. J.: Hydrogen and helium ions. Ann. Geophys. **22**, 247—251 (1966).

120. Walker, J. C. G.: Thermal diffusion in the topside ionosphere. Planetary Space Sci. **15**, 1151—1156 (1967).

121. Schunk, R., Walker, J. C. G.: Thermal diffusion in the F_2 region of the ionosphere. Planetary Space Sci. **18**, 535—557 (1970).

122. Schunk, R., Walker, J. C. G.: Thermal diffusion in the topside ionosphere for mixtures which include multiply-charged ions. Planetary Space Sci. **17**, 853—868 (1969).

123. Chandra, S., Goldberg, R. A.: Geomagnetic control of diffusion in the upper atmosphere. J. Geophys. Res. **69**, 3187 (1964).

124. Angerami, J. J., Thomas, J. O.: Studies of planetary atmospheres, 1, The distribution of electrons and ions in the Earth's exosphere. J. Geophys. Res. **69**, 4537 (1964).

125. Obayashi, T., Maeda, K. I.: The electrical state of the upper atmosphere. Problems of atmospheric and space electricity, S. C. Coroniti (ed.), pp. 532—547. Elsevier Publ. Co. 1965.

126. Rishbeth, H.: The F layer dynamo. Planetary Space Sci. **19**, 263—267 (1971).

127. Rishbeth, H.: Thermospheric winds and the F region: A review. J. Atmospheric Terrest. Phys. **34**, 1—47 (1972).

128. Eviatar, A., Lenchek, A. M., Singer, S. F.: Distribution of density in an ion-exosphere of a non-rotating planet. Phys. Fluids 7, 1775—1779 (1964).

129. Hartle, R. E.: Ion-exosphere with variable conditions at the baropause. Phys. Fluids **12**, 455 (1969).

130. Ness, N. F.: Observations of the interaction of the solar wind with the geomagnetic field during quiet conditions. Ch. III (57—89). Solar-terrestrial Physics, J. W. King, W. S. Newman, eds. N. Y.: Academic Press 1967.

131. Donahue, T. M.: Polar ion flow: Wind or breeze?, Rev. Geophys. **9**, 1—10 (1971).

132. Dessler, A. J., Michel, F. C.: Plasma in the geomagnetic tail. J. Geophys. Res. **71**, 1421—1426 (1966).

133. Nishida, A.: Formation of a plasmapause, or magnetospheric plasma knee by combined action of magnetospheric convection and plasma escape from the tail. J. Geophys. Res. **71**, 5669—5680 (1966).

134. Banks, P. M., Holzer, T. E.: The polar wind. J. Geophys. Res. **73**, 6855—6868 (1968).

135. Banks, P. M., Holzer, T. E.: High latitude plasma transport: the polar wind. J. Geophys. Res. **74**, 6317—6332 (1969).

136. Marubashi, K.: Escape of the polar-ionospheric plasma into the magnetospheric tail, Rept. Ion. Space Res. Japan **24**, 322—346 (1970).

137. Lemaire, J., Scherer, M.: Model of the polar ion-exosphere. Planetary Space Sci. **18**, 103—130 (1970).

138. Rishbeth, H.: A review of ionospheric F region theory. Proc. IEEE **55**, 16—35 (1967).

139. Chandra, S.: Electron density distribution on the upper F region. J. Geophys. Res. **68**, 1937—1942 (1963).

140. Rishbeth, H.: The effects of winds on the ionospheric F_2 peak. J. Atmospheric Terrest. Phys. **29**, 225—238 (1967).

141. Hanson, W. B.: Upper atmosphere helium ions. J. Geophys. Res. **67**, 183—188 (1962).

142. Wright, J. W.: A model of the F-region above $h_{max}F_2$. J. Geophys. Res. **65**, 185—191 (1960).

143. Chandra, S., Herman, J. R.: The influence of varying solar flux on ionopheric temperatures and densities: A theoretical study. Planetary Space Sci. **17**, 815—840 (1969).

144. Stubbe, P.: Simultaneous solution of the time-dependent coupled continuity equations, heat conduction equations, and equations of motion for a system consisting of a neutral gas, an electron gas and a four-component ion gas. J. Atmospheric Terrest. Phys. **32**, 865—903 (1970).

145. Nisbet, J. S.: On the construction and use of a simple ionospheric model. Radio Sci. **6**, 437—464 (1971).

146. Hess, W. N.: The Radiation Belt and Magnetosphere. Waltham, Mass.: Blaisdell Publ. Co. 1968.

147. Carpenter, D. L.: Whistler studies of the plasmapause in the magnetosphere —1: Temporal variation in the position of the knee and some evidence on plasma motions near the knee. J. Geophys. Res. **71**, 693 (1966).

147a. Carpenter, D. L., Park, C. G.: On what ionospheric workers should know about the plasmapause-plasmasphere. Rev. Geophys. **11**, 133—154 (1973).

148. Axford, W. I.: Magnetospheric convection. Rev. Geophys. **7**, 421 (1969).

149. Brice, W. M.: Bulk motion of the magnetosphere. J. Geophys. Res. **72**, 5193 (1967).

150. Brice, W. M., Ioannides, G. A.: The magnetospheres of Jupiter and Earth. Icarus **13**, 173—183 (1970).

151. Ioannides, G., Brice, N.: Plasma densities in the Jovian Magnetosphere: Plasma sling-shot or Maxwell demon. Icarus **14**, 360—373 (1971).

151a. Dolginov, Sh. Sh., Yeroshenko, Ye. G., Zhuzgov, L. N.: The magnetic field in the very close neighborhood of Mars according to data from the Mars 2 and 3 spacecraft. J. Geophys. Res. **78**, July (1973).

151b. Bauer, S. J., Hartle, R. E.: On the extent of the Martian ionosphere. J. Geophys. Res. **78**, 3169—3171 (1973).

152. Spreiter, J. R., Summers, A. L., Rizzi, A. W.: Solar wind flow past non-magnetic planets; Mars and Venus. Planetary Space Sci. **18**, 1281 (1970).

152a. Michel, F. C.: Solar wind interaction with planetary atmospheres. Rev. Geophys. **9**, 427—436 (1971).

153. Cloutier, R. A., McElroy, M. B., Michel, F. C.: Modification of the Martian inosphere by the solar wind. J. Geophys. Res. **74**, 6215 (1969).

154. Michel, F. C.: Solar-wind induced mass loss from magnetic field-free planets. Planetary Space Sci. **19**, 1580—1583 (1971).

155. Wallis, M.: Shock-free deceleration of the solar wind. Nature **233**, 23—25 (1971).

155a. Wallis, M.: Comet-like interaction of Venus with the solar wind. I, Cosmic Electrodynamics **3**, 4559 (1972).

156. Kunkel, W. B. (ed.): Plasma Physics in Theory and Application. New York, N. Y.: McGraw-Hill Book Co. 1966.

156a. Yeh, K. C., Liu, C. H.: Propagation and Application of waves in the ionosphere. Rev. Geophys. **10**, 631—709 (1972).

157. Stix, T. H.: The Theory of Plasma Waves. New York: McGraw-Hill 1962.

158. Astrom, E. O.: On waves in an ionized gas. Arkiv. Fysik. **2**, 443 (1950).

159. Ratcliffe, J. A.: The Magneto-Ionic Theory and its Applications to the Ionosphere. Cambridge: University Press 1969.

160. Davies, K.: Ionospheric Radio Propagation. NBS Monograph 80, US GPO 1965.

161. Helliwell, R. A.: Whistlers and Related Ionospheric Phenomena. Stanford University Press 1965.

162. Bekefi, G.: Radiation Processes in Plasmas. New York: John Wiley & Sons 1966.

163. Dougherty, J. P., Watson, S. R.: The interpretation of plasma resonances observed by ionospheric topside sounders. Adv. Plasma Phys. Vol. 4, pp. 1—41. H. Simon & W. B. Tompson, eds. John Wiley & Sons, Inc. 1971.

164. Oya, H.: Verification of theory of weak turbelence relating to the sequence of diffuse plasma resonances in space. Phys. Fluids **14**, 2487—2499 (1971).

165. Alfvén, H., Fälthammar, C. G.: Cosmical Electrodynamics. Oxford: University Press 1963.

166. Jacobs, J. A.: Geomagnetic Micropulsations. Berlin-Heidelberg-New York: Springer 1970.

167. Dessler, A. J.: Ionospheric heating by hydromagnetic waves. J. Geophys. Res. **64**, 397—401 (1959).

168. Hasegawa, A.: Plasma instabilities in the magnetosphere. Rev. Geophys. **9**, 703—772 (1971).

169. Thompson, W. B.: An Introduction to Plasma Physics. London: Pergamon Press 1962.

170. Banks, P. M.: Difficulties with thermal protons in the Venusian topside ionosphere. J. Geophys. Res. **76**, 8454—8456 (1971).

171. Gold, T.: Motions in the magnetosphere of earth. J. Geophys. Res. **64**, 1219—1224 (1959).

172. Scarf, F. L.: Plasma in the magnetosphere. Adv. Plasma Phys. Vol. **1**, A. Simon and W. B. Thompson, eds. pp. 101—152, New York: John Wiley 1968.

173. Farley, D. T.: A plasma instability resulting in field aligned irregularities in the ionosphere. J. Geophys. Res. **68**, 6083 (1963).

173a. Farley, D. T., Balsley, B. B.: Instabilities in the equatorial electrojet. J. Geophys. Res. **78**, 227—239 (1973).

173b. Perkins, F.: Spread F and ionospheric currents. J. Geophys. Res. **78**, 218—226 (1973).

174. Rawer, K., Suchy, K.: Radio observations of the ionosphere. Handb. Phys. Vol. **49**/2 Geophys. III/2. pp. 1—546. Berlin-Heidelberg-New York: Springer 1967.

175. Garriott, O. K., da Rosa, A. V., Ross, W. J.: Electron content obtained from Faraday rotation and phase pathlength variations. J. Atmospheric Terrest. Phys. **32**, 705—727 (1970).

176. Fjeldbo, G., Eshleman, V. R., Garriott, O. K., Smith, F. L. III: The two frequency bistatic radar occultation method for the study of planetary ionospheres. J. Geophys. Res. **70**, 3701—3710 (1965).

177. Fjeldbo, G., Eshleman, V. R.: The bistatic radar-occultation method for the study of planetary atmospheres. J. Geophys. Res. **70**, 3217—3225 (1965).

178. Pirraglia, J., Gross, S. H.: Latitudinal and longitudinal variation of a planetary atmosphere and the occultation experiment. Planetary Space Sci. **18**, 1769—1784 (1970).

179. Hagg, E. L., Hewens, E. J., Nelms, G. L.: The interpretation of topside sounder ionograms. Proc. IEEE **57**, 949—959 (1969).

180. Jackson, J. E.: The reduction of topside ionograms to electron density profiles. Proc. IEEE **57**, 960—976 (1969).

181. Jackson, J. E.: The p'(f) to N(h) inversion problem in ionospheric soundings. Proc. Workshop, Mathematics of Profile Inversion, 4-15—4-26, L. Colin, ed. NASA TMX-62, 150, 1972.

182. Calvert, W., McAfee, J. R.: Topside sounder resonances. Proc. IEEE **57**, 1097—1107 (1969).

183. Gordon, W. E.: Incoherent scattering by free electrons with applications to space exploration by radar. Proc. IRE, **46**, 1824—1829 (1958).

184. Bowles, K. L.: Observations of vertical incidence scatter from the ionosphere at 41 Mc/s. Phys. Rev. Letters **1**, 454—455 (1958).

185. Evans, J. V.: Theory and Practice of ionosphere study by Thomson scatter radar. Proc. IEEE **57**, 495—530 (1969).

186. Evans, J. V.: Ionospheric movements measured by incoherent scatter—a review. J. Atmospheric Terrest. Phys. **34**, 175—209 (1972).

187. Cohen, R., Utlaut, W. F.: Modifying the ionosphere with intense radio waves. Science **174**, 245 (1971).

188. Boyd, R. L. F.: An introduction to Langmuir probes for space research. Ch. 38 (pp. 455—465). Solar-Terrestrial Relations, Ortner and Maseland, eds. New York: Gordon and Breach, Sci. Publ. 1966.

188a. Clark, D. H., Raitt, W. J., Willmore, A. P.: A measured anisotropy in the ionospheric electron temperature. J. Atm. Terr. Phys. **35**, 63—76 (1973).

189. Bourdeau, R. E.: Ionospheric research from space vehicles. Space Sci. Rev. **1**, 683 (1961).

190. Kiser, R. W.: Introduction to Mass Spectrometry and its Applications. Prentice Hall Inc. 1965.

191. Spencer, N. W.: Upper atmosphere studies by mass spectrometry. Advances in Mass-Spectrometry, Vol. **5**, pp. 509—519. London: The Inst. of. Petroleum 1971.

192. Jackson, J. E., Kane, J. A.: Measurement of ionospheric electron densities using a rf probe technique. J. Geophys. Res. **64**, 8 (1959).
193. Hartz, T. R.: Radio noise levels within and above the ionosphere. Proc. IEEE **57**, 1042—1051 (1969).
194. Bauer, S. J., Stone, R. G.: Satellite observations of radio noise in the magnetosphere. Nature **218**, 1145—1147 (1968).
195. Barrington, R. E.: Ionospheric ion composition deduced from VLF observations. Proc. IEEE **57**, 1036—1041 (1969).
196. Warnock, J. M., McAfee, J. R., Thompson, T. L.: Electron temperature from topside plasma resonance observations. J. Geophys. Res. **75**, 7272—7275 (1970).
197. Oya, H., Benson, R. F.: A new method for in-situ electron temperature determinations from plasma wave phenomena. J. Geophys. Res. **77**, 4272—4276 (1972).
198. Hoegy, W. R.: Probe and radar electron temperatures in an isotropic non-equilibrium plasma. J. Geophys. Res. **76**, 8333—8340 (1971).
199. Thomas, L.: The lower ionosphere. J. Atmospheric Terrest. Phys. **33**, 157—195 (1971).
200. Aikin, A. C.: Ionization sources of the ionospheric D and E regions. COSPAR Symposium on D and E region. Urbana, Ill.: Ion chemistry 1971.
201. Narcisi, R. S.: On water cluster ions in the ionospheric D region. Planetary Electrodynamics, Vol. **2**, p. 69. Edited by S. C. Coroniti and J. Hughes. New York: Gordon and Breach 1969.
202. Goldberg, R. A., Aikin, A. C.: Studies of positive ion composition in the equational D-region ionosphere. J. Geophys. Res. **76**, 8352—8364 (1971).
203. Reid, G. C.: Production and loss of electrons in the quiet daytime D region of the ionosphere. J. Geophys. Res. **75**, 2551 (1970).
204. LeLevier, R. E., Branscomb, L. M.: Ion chemistry governing mesospheric electron concentrations. J. Geophys. Res. **73**, 27 (1968).
205. Arnold, F., Krankowsky, D.: Negative ions in the lower ionosphere: A comparison of a model calculation and a mass-spectrometric measurement. J. Atmospheric Terrest. Phys. **33**, 1693—1702 (1971).
206. Goldberg, R. A., Blumle, L. J.: Positive composition from a rocket-borne mass spectrometer. J. Geophys. Res. **75**, 133—142 (1970).
207. Whitehead, J. D.: Report on the production and prediction of sporadic E. Rev. Geophys. **8**, 65 (1970).
208. Rishbeth, H.: On explaining the behaviour of the ionospheric F region. Rev. Geophys. **6**, 33—71 (1968).
209. Kohl, H., King, J. W.: Atmospheric winds between 100 and 700 km and their effects on the ionosphere. J. Atmospheric Terrest. Phys. **29**, 1045—1062 (1967).
210. Rishbeth, H.: Thermospheric winds and the F region: A review. J. Atmospheric Terrest. Phys. **34**, 1—47 (1972).
211. Strobel, D. F., McElroy, M. B.: The F_2 layer at middle latitudes. Planetary Space Sci. **18**, 1181—1202 (1970).
211a. Becker, W.: The standard profile of the mid-latitude F-region of the ionosphere as deduced from bottomside and topside ionograms. Space Research **XII**, pp. 1241—1252. Berlin: Akademie-Verlag 1972.
212. Wright, J. W.: Dependence of the ionospheric F region on the solar cycle. Nature **194**, 462 (1962).
213. Mayr, H. G., Mahajan, K. K.: Seasonal variation in the F_2 region. J. Geophys. Res. **76**, 1017—1027 (1971).

214. Walker, J. C. G., Spencer, N. W.: Temperature of the earth's upper atmosphere. Science **162**, 1437—1442 (1968).
215. Evans, J. V.: Midlatitude electron and ion temperatures at sunspot minimum. Planetary Space Sci. **15**, 1557—1570 (1967).
216. Brace, L.: The global structure of ionosphere temperature. Space Research **X**, p. 633, Amsterdam: North Holland Publ. Co. 1970.
217. Sanatani, S., Hanson, W. B.: Plasma temperatures in the magnetosphere. J. Geophys. Res. **75**, 769—775 (1970).
218. Willmore, A. P.: Electron and ion temperatures in the ionosphere. Space Sci. Rev. **11**, 607 (1970).
219. Sterling, D. L., Hanson, W. B., Moffett, R. J., Baxter, R. G.: Influence of electromagnetic drifts and neutral air winds on some features of the F_2 region. Radio Sci. **4**, 1005—1023 (1969).
220. Thomas, L.: The F_2 region equatorial anomaly during solstice periods at sunspot maximum. J. Atmospheric Terrest. Phys. **30**, 1631—1640 (1968).
221. Rush, C. M., Rush, S. V., Lyons, L. R., Venkateswaran, S. V.: Equatorial anomaly during a period of declining solar activity. Radio Sci. **4**, 829—841 (1969).
222. Rishbeth, H.: The polar F region. The Polar Ionosphere and Magnetospheric Processes, pp. 175—192. G. Svocli (ed.), New York: Gordon and Breach Science Publishers 1970.
223. Andrews, M. K., Thomas, J. O.: Electron density distribution above the winter pole. Nature **221**, 223-227 (1969).
224. Muldrew, D. B.: F-layer ionization troughs deduced from Alouette data. J. Geophys. Res. **70**, 2635—2650 (1965).
225. Taylor, H. A.: The light ion trough. Planetary Space Sci. **20**, 1593 (1972).
225a. Taylor, H. A., Walsh, W. J.: The light ion trough, the main trough and the plasmapause. J. Geophys. Res. **77**, 6716—6723 (1972).
226. Brinton, H. C., Grebowsky, J. M., Mayr, H. G.: Altitude variation of ion composition in the midlatitude trough region: Evidence for an upward plasma flow. J. Geophys. Res. **76**, 3739—3475 (1971).
227. Evans, J. V.: F-region heating observed during the main phase of magnetic storms. J. Geophys. Res. **75**, 4815—4823 (1970).
228. Chandra, S., Stubbe, P.: Ion and neutral composition changes in the thermospheric region during magnetic storms. Planetary Space Sci. **19**, 491—502 (1971).
228a. Thomas, L.: F_2 region disturbances associated with major magnetic storms. Planetary Space Sci. **18**, 917—928 (1970).
229. Bauer, S. J., Krishnamurthy, B. V.: Behavior of the topside ionospheric during a great magnetic storm. Planetary Space Sci. **16**, 653—663 (1968).
229a. Obayashi, T., Matuura, N.: Theoretical model of F-region storms. Solar-Terrestrial Physics, 1970, Part IV, E. Dyer (ed.), Reidel Publishing Co., 119—211, 1972.
230. Nagy, A. F., Roble, G. R., Hays, P. B.: Stable mid-latitude red arcs: Observations and theory. Space Sci. Rev. **11**, 709 (1970).
231. Chandra, S., Maier, E. J., Stubbe, P.: The upper atmosphere as a regulator of sub-auroral red arcs. Planetary Space Sci. **20**, 461—472 (1972).
232. Herman, J. R.: Spread F and ionospheric F region irregularities. Rev. Geophys. **4**, 255 (1966).
233. Farley, D. T., Balsley, B. B., Woodman, R. F., McClure, J. P.: Equatorial spread F: Implications of VHF radar observations. J. Geophys. Res. **75**, 7199—7216 (1970).

234. Hanson, W. B., Sanatani, S.: Relationship between Fe^+ ions and equatorial spread F. J. Geophys. Res. **76**, 7761—7768 (1971).
235. Fjeldbo, G., Eshleman, V. R.: The atmosphere of Mars analyzed by integral inversion of the Mariner 4 occultation data. Planetary Space Sci. **16**, 1035—1059 (1968).
236. Hunten, D. M.: The ionosphere and upper atmosphere of Mars. The Atmospheres of Venus and Mars. J. C. Brandt and M. B. McElroy (eds.), New York: Gordon and Breach 1968.
236a. Gringauz, K. I., Breus, T.: Comparative characteristics of the ionospheres of the planets of the terrestrial group: Mars, Venus and the Earth. Space Sci. Rev. **10**, 743—769 (1970).
237. Rasool, S. I., Stewart, R. W.: Results and interpretation of S-band occultation experiments on Mars and Venus. J. Atmospheric Sci. **28**, 869—878 (1971).
238. Kliore, A. J., Cain, D. L., Fjeldbo, G., Weidel, B. L., Rasool, S. I.: Mariner 9 S-band occultation experiment: Initial results on the atmosphere and topography of Mars. Science **175**, 313—317 (1972).
239. Aikin, A. C.: The lower ionosphere of Mars. Icarus **9**, 487—497 (1968).
240. Whitten, R. C., Poppoff, R. C., Sims, J. S.: The ionosphere of Mars below 80 km altitude. I. Planetary Space Sci. **19**, 243—250 (1971).
241. Barth, C. A., Hord, C. W., Stewart, A. I., Lane, A. L.: Mariner 9 ultraviolet spectrometer experiment initial results. Science **175**, 309—312 (1972).
242. Kliore, A. J., Fjeldbo, G., Seidel, B. L., Sykes, J. J., Woiceshyn, P. M.: S-band radio occultation measurements of the atmosphere and topography of Mars with Mariner 9—extended mission coverage of polar und intermediate latitudes. J. Geophys. Res. **78**, July (1973).
243. McElroy, M. B., Connell, J. C.: Atomic carbon in the atmospheres of Mars and Venus. J. Geophys. Res. **76**, 6674—6690 (1971).
244. Stewart, A. I.: Mariner 6 and 7 ultraviolet spectrometer experiment: Implications of CO_2^+, CO and O airglow. J. Geophys. Res. **77**, 54—68 (1972).
245. Stewart, R. W., Hogan, J. S.: Empirical determination of heating efficiencies in the Mars and Venus atmospheres. J. Atmospheric Sci., **26**, 330 (1969).
246. Stewart, R. W.: The electron distributions in the Mars and Venus upper atmospheres. J. Atmospheric Sci. **28**, 1069—1073 (1971).
247. Eshleman, V. R.: Atmospheres of Mars and Venus: A review of Mariner 4 and 5 and Venera 4 experiments. Radio Sci. **5**, 325—332 (1970).
248. McElroy, M. B.: Structure of the Venus and Mars Atmospheres. J. Geophys. Res. **74**, 29 (1969).
249. Whitten, R. C.: The daytime upper ionosphere of Venus. J. Geophys. Res. **75**, 3707 (1970).
250. Bauer, S. J., Hartle, R. E., Herman, J. R.: Topside ionosphere of Venus and its interaction with the solar wind. Nature **225**, 533 (1970).
251. Barth, C. A., Wallace, L., Pearce, J. B.: Mariner 5 measurements of Lyman-alpha radiation near Venus. J. Geophys. Res. **73**, 2541—2545 (1968).
252. Kurt, V. G., Dostovalow, S. B., Sheffer, E. K.: The Venus far ultraviolet observations with Venera 4. The Venus Atmosphere. R. Jastrow and S. I. Rasool (ed.), pp. 377—485. N. Y.: Gordon and Breach 1969.
253. Banks, P. M., Axford, W. I.: Origin and dynamical behavior of thermal protons in the Venusian ionosphere. Nature **225**, 924 (1970).
254. McElroy, M. B., Strobel, D. F.: Models for the nighttime Venus ionosphere. J. Geophys. Res. **74**, 1118—1127 (1969).
255. Hartle, R. E., Herman, J. R., private communication.

256. Aikin, A. C.: Ion clusters and the Venus ultraviolet haze layer, Nature **235**, 10—11 (1972).
257. Warwick, J. W.: Radiophysics of Jupiter. Space Sci. Rev. **6**, 841—891 (1967).
258. Gross, S. H., Rasool, S. I.: The upper atmosphere of Jupiter. Icarus **3**, 311 (1964).
259. Hunten, D. M.: The upper atmosphere of Jupiter. J. Atmospheric Sci. **26**, 827 (1969).
260. Prasad, S. S., Capone, L. A.: The Jovian Ionosphere: Composition and temperature. Icarus **15**, 45—55 (1971).

List of Symbols

A	constant; area
a	constant, factor
$\boldsymbol{B}\ (B)$	magnetic field vector (magnitude)
$\mathscr{B}\ B$	constant
b	constant
C	capacity (heat, electric)
Ch( )	Chapman function
c	velocity of light in vacuum
c_i	ion acoustic speed
c_s	speed of sound
D	distance; dispersion relation parameter
D_a	ambipolar (plasma) diffusion coefficient
D_B	Bohm (turbulent plasma) diffusion coefficient
D_j	molecular diffusion coefficient
d; d	diameter; day
$\boldsymbol{E}\ (E)$	electric field
E	energy
e; e	base of natural logarithm; electron charge (esu)
$\boldsymbol{F}$	force
F_j	particle flux
$\mathscr{F}$	heat flux
f; $\mathscr{f}$	wave frequency; oscillator strength
$f(\ )$	distribution function
G	gravitational constant
g	acceleration of gravity
$\mathscr{g}$	emission rate factor
H	atmospheric (pressure) scale height
H_ρ	density scale height
H_N	plasma density scale height
$\mathscr{H}$	plasma (pressure) scale height
h; $\mathscr{h}$	altitude; Planck's constant
I; I	intensity of radiation; magnetic dip (inclination)
$\mathscr{I}$	column emission rate (airglow)
i	imaginary number $\sqrt{-1}$

J	ionization rate coefficient
$J(>P)$	integral rigidity spectrum
$j(E)$	differential energy spectrum
$\boldsymbol{j}\ (j)$	electric current
$(\boldsymbol{K})$	dielectric tensor
$K(T)$	heat conductivity
K_D	eddy diffusion coefficient
$\boldsymbol{k}, k$	wave vector, propagation constant
k	rate constant
$\mathscr{k}$	Boltzmann constant
L	scale length; loss rate; McIlwain parameter; dispersion relation parameter
$\mathscr{M}$	magnetic moment $(R_0^3 B_s)$
M	mass; Mach number
m	mean molecular mass
m_j	mass of jth constituent
$\hat{m}_j$	mass in AMU
m_+	mean ion mass
N	plasma number density
$\tilde{N}$	refractivity
$\mathscr{N}$	total content (number of particles in a column of $1\ \mathrm{cm}^2$ cross section)
n	harmonic number
n	neutral particle number density; refractive index
P	rigidity; dispersion relation parameter; phase path length; probability.
P'	group path length
p	pressure
Q	heating rate
q	ion-pair production (ionization) rate; charge density (eN)
R	planetocentric distance; dispersion relation parameter; geometric range
R_0	planetary radius
R^*	minimum radial distance of grazing ray
$\mathscr{R}$	particle range $(\mathrm{g\ cm}^{-2})$
r	radius vector
r_B	Larmor (cyclotron, gyro) radius
S	surface; flux; dispersion relation parameter
$S_{10.7}$	10.7 cm (2800 MHz) solar radio flux
s	pathlength
T	absolute temperature
t	time
u_0	most probable velocity

$\mathscr{V}$; V volume; electric potential (voltage)
V_A Alfvén velocity
V_1, V_2 hydromagnetic wave mode velocities
$\boldsymbol{v}$, (v) particle or wave velocity
$\boldsymbol{v}_D$ plasma (ambipolar) diffusion velocity
$\boldsymbol{v}_E$ plasma drift velocity
$\boldsymbol{v}_g$ group velocity
$\boldsymbol{v}_{ph}$ phase velocity
v_l longitudinal electron velocity $(\mathscr{k}T_e/m_e)^{\frac{1}{2}}$
v_∞ escape velocity
$\langle v_R \rangle$ effusion velocity
W energy
$\boldsymbol{w}_j$ molecular diffusion velocity
w wind speed
X escape parameter $(v_\infty/u_0)^2$; magnetoionic parameter; atomic constituent
x pathlength; cartesian coordinate; Chapman function parameter (R/H)
Y rotational (escape) parameter $(\Omega R/u_0)$; magnetoionic parameter; atomic constituent
y cartesian coordinate
Z normalized height parameter (z/H); magnetoionic parameter; charge number; atomic constituent
z height parameter; cartesian coordinate.

α polarizability; pitch angle; factor of proportionality; meteor line density; incoherent scatter parameter
α_D dissociative recombination coefficient
α_j thermal diffusion factor
α_{mn} ion-ion recombination (mutual neutralization) coefficient
α_r radiative recombination coefficient
β scale height gradient (dH/dz); linear loss rate coefficient
$\Gamma_\infty()$ complete gamma function
$\Gamma_\kappa()$ incomplete gamma function
γ ratio of specific heats (adiabatic index); growth rate; Euler's constant
δ mass density (meteor particles), declination (solar)
ε_j heating efficiency
ζ height parameter
η viscosity
η_i ionization efficiency (σ_i/σ_a)
η_{jk} relative abundance
θ angle between wave vector $\boldsymbol{k}$ and magnetic field $\boldsymbol{B}$

222

θ_C Čerenkov cone halfangle
κ attenuation factor
Λ Coulomb parameter
λ wavelength; mean free path
λ^- negative ion-electron concentration ratio
λ_D Debye length
μ_{jk} reduced mass $(m_j m_k / m_j + m_k)$
ν frequency
ν_{jk} collision frequency between jth and kth constituent
ξ partition function
Π_k generalized plasma frequency (dispersion relation)
ρ mass density
σ cross section (general)
σ_D momentum cross section
(σ) conductivity tensor
τ time constant; period
$\Phi(\ )$ error integral
Φ ionizing radiation flux
ϕ potential; particle flux
φ latitude
χ zenith angle
Ψ arbitrary variable
ψ aspect angle
Ω solid angle; rotation rate; angular frequency (ion modes)
ω angular frequency

Subscripts

B gyro-, cyclotron-
b boundary
C chemical
c critical
D diffusion; dissociation
$e; E$ electron; electric field
g gravitational
i ion; ionization
j jth constituent
k kth constituent
l longitudinal, i.e., parallel to wave vector k
M Maxwellian
m maximum, peak
N plasma
n neutral

o; *o*	reference level; ordinary mode
P	planet
p	proton; plasma
R, *r*	radial
S	surface
s	satellite, spacecraft
w	wind
x	*x*-component; extraordinary mode
y	*y*-component
z	*z*-component; *z*-mode
‖	component parallel to **B**
⊥	component perpendicular to **B**
∞	upper limit; escape
⊙	Sun, solar
⊕	Earth
♂	Mars
♀	Venus
♃	Jupiter
+	(positive) ion
−	negative ion

Subject Index

Absorption cross sections 53—55
—, radio wave 137, 180
Acoustic velocity 16, 153
— waves 17, 81
Adiabatic lapse rate 3
Age of planets 30
Air drag 106
Airglow 48, 93—95
—, column emission rate 49, 50, 94
— emissions 57, 72
Alfvén velocity 125, 146
— wave 146
Altitude, potential (reduced) 8
— of unit optical depth (see also penetration depth) 51, 52, 120, 121
Anomaly, diurnal phase 20—22
—, geomagnetic 189
—, seasonal 188
Appleton-Hartree formula 136, 137
Atmospheres, evolution of 40
—, Jovian planets 43, 44, 205
—, nomenclature 3—6
—, physical properties 40—44, 205
—, terrestrial planets 40—42, 205
—, total content 7
—, total mass 7
Atmospheric blow-off 35
— escape, see escape
Attachment law 91
— processes 90, 91
Auroral oval 190, 192
— particles 61
— zone 184

Backscatter, see incoherent scatter
Baropause 5, 27, 108
Barosphere 5, 6—17
Barometric law 7, 35
Bennett rf mass spectrometer 174, 186
Bohm diffusion 76
Bohm-Gross (warm plasma) correction 141, 166

Boltzmann equation, collisionless 36, 134
Bow shock 125, 128, 129
Bradbury layer 114, 116, 117, 121, 187
Breathing velocity, atmospheric 18
Brunt-Väisälä frequency 17
Butler and Buckingham formula 71, 72

C layer 57, 180
Capacity, thermal (atmosphere) 22
Čerenkov mechanism 71, 142, 143, 166
Chapman function 20, 49, 55, 56, 159
— ion production function 52
— layer 113, 115, 120, 121, 130, 180, 183, 185, 194, 197
— — properties 113—116, 119, 197
Charge transfer reactions 86—90
— neutrality 92, 101
Charge transfer 87—89
— —, dissociative 87
— —, resonant 88
Cluster ions 90, 131, 180, 181, 202
CMA diagram 140
Cold plasma approximation 136
— — dispersion relation, see also Appleton-Hartree formula 136
Collisions, Coulomb 68, 74, 132, 134
—, elastic 67, 72, 73
—, inelastic 67, 72, 73
Collision frequency, Coulomb (electron-ion) 68
— —, electron-neutral 73
— —, ion-neutral 74, 97
— time 68, 104, 106, 107
Conductivity, electrical 105
— tensor 105
Conductivity tensor, components 105, 106
—, thermal, see thermal conductivity

Physics and Chemistry in Space

A series of monographs written
and published to serve
the student, the teacher and
the researcher with a clear and
concise presentation of up-to-date
topics of space exploration.

Edited by J. G. Roederer, Denver, Colo.

Editorial Board: H. Elsässer, Heidelberg; G. Elwert, Tübingen;
L. G. Jacchia, Cambridge, Mass.; J. A. Jacobs, Edmonton, Alta.;
N. F. Ness, Greenbelt; Md.; W. Riedler, Graz

Volume 5
Hundhausen: Coronal Expansion and Solar Wind

By A. J. Hundhausen,
High Altitude Observatory,
National Center for Atmospheric
Research, Boulder, Colo. USA
With 101 figs. XII, 238 pp. 1972
Cloth DM 68,—; US $30.70

The author gives a physical interpretation of basic solar wind phenomena, based on a synthesis of interplanetary observations and theoretical
models of the coronal expansion.

Springer-Verlag
Berlin
Heidelberg
New York

London München Paris
Sydney Tokyo Wien

Volume 1
Jacobs: Geomagnetic Micropulsations

By J. A. Jacobs, Killam Memorial,
Professor of Science, The University
of Alberta, Edmonton, Canada
With 81 figs. VIII, 179 pp. 1970
Cloth DM 36,—; US $16.30

A detailed account both of the
morphology of geomagnetic
micropulsations and of the various
theories that have been proposed
to explain them.

Volume 3
Adler/Trombka: Geochemical Exploration of the Moon and Planets

By Dr. I. Adler, Senior Scientist,
and Dr. J. I. Trombka,
both: Goddard Space Flight Center,
NASA, Greenbelt, Md, USA
With 129 figs. X, 243 pp. 1970
Cloth DM 58,—; US $26.20

A review of progress in the
geochemical exploration
of the Moon and planets and
of future plans for lunar and
planetary exploration.

Volume 2
Roederer: Dynamics of Geomagnetically Trapped Radiation

By J. G. Roederer, Professor of
Physics, University of Denver,
Denver, Colo., USA
With 94 figs. XIV, 166 pp. 1970
Cloth DM 36,—; US $16.30

A concise, systematic and up-to-
date discussion of the basic
dynamical processes governing the
earth's radiation belts, with guide-
lines for quantitative applications
of the theory.

Volume 4
Omholt: The Optical Aurora

By A. Omholt, Professor of Physics,
Universitetet i Oslo, Fysisk
Institutt, Blindern, Oslo, Norway
With 54 figs. XIII, 198 pp. 1971
Cloth DM 58,—; US $26.20

This book deals with the optical
aurora, its occurrence and properties and the way it is produced by
the primary electrons and protons.
The auroral spectrum and its
excitation is treated in great detail.

Prices are subject to change
without notice

A Publication of the Astronomisches Rechen-Institut Heidelberg, Member of the Abstracting Board of the International Council of Scientific Unions. Edited by S. Böhme, W. Fricke, U. Güntzel-Lingner, F. Henn, D. Krahn, U. Scheffer, G. Zech

ASTRONOMY AND ASTROPHYSICS ABSTRACTS

Astronomy and Astrophysics Abstracts is prepared under the auspices of the International Astronomical Union
Two volumes are scheduled to appear per year

Published for the Astronomisches Rechen-Institut by

SPRINGER-VERLAG BERLIN HEIDELBERG NEW YORK

München · London · Paris Sydney · Tokyo · Wien

Vol. 1
Literature 1969, Part 1
VIII, 435 pages. 1969

Vol. 2
Literature 1969, Part 2
X, 516 pages. 1970

Vol. 3
Literature 1970, Part 1
X, 490 pages. 1970

Vol. 4
Literature 1970, Part 2
X, 562 pages. 1971

Vol. 5
Literature 1971, Part 1
X, 505 pages. 1971

Vol. 6
Literature 1971, Part 2
X, 560 pages. 1972

Vol. 7
Literature 1972, Part 1
X, 526 pages. 1972

Price per volume:
Cloth DM 72,—
US $32.50

Subscription price per volume: Cloth DM 57,60
US $26.00

Subscription may still be started from Volume 1 at the subscription price.

Subscriptions should be addressed to one of the following:

- Springer-Verlag, 1 Berlin 33, Heidelberger Platz 3 (Germany)
- Astronomisches Rechen-Institut, 69 Heidelberg, Mönchhofstrasse 12-14 (Germany)
- Springer-Verlag New York Inc. 175 Fifth Avenue New York, N.Y. 10010 (USA)
- your local bookseller

Prices are subject to change without notice